SpringerBriefs in Molecular Science

Chemistry of Foods

Series Editor

Salvatore Parisi, Lourdes Matha Institute of Hotel Management, Kerala, India

The series Springer Briefs in Molecular Science: Chemistry of Foods presents compact topical volumes in the area of food chemistry. The series has a clear focus on the chemistry and chemical aspects of foods, topics such as the physics or biology of foods are not part of its scope. The Briefs volumes in the series aim at presenting chemical background information or an introduction and clear-cut overview on the chemistry related to specific topics in this area. Typical topics thus include:

- Compound classes in foods—their chemistry and properties with respect to the foods (e.g. sugars, proteins, fats, minerals, …)
- Contaminants and additives in foods—their chemistry and chemical transformations
- Chemical analysis and monitoring of foods
- Chemical transformations in foods, evolution and alterations of chemicals in foods, interactions between food and its packaging materials, chemical aspects of the food production processes
- Chemistry and the food industry—from safety protocols to modern food production

The treated subjects will particularly appeal to professionals and researchers concerned with food chemistry. Many volume topics address professionals and current problems in the food industry, but will also be interesting for readers generally concerned with the chemistry of foods. With the unique format and character of SpringerBriefs (50 to 125 pages), the volumes are compact and easily digestible. Briefs allow authors to present their ideas and readers to absorb them with minimal time investment. Briefs will be published as part of Springer's eBook collection, with millions of users worldwide. In addition, Briefs will be available for individual print and electronic purchase. Briefs are characterized by fast, global electronic dissemination, standard publishing contracts, easy-to-use manuscript preparation and formatting guidelines, and expedited production schedules.

Both solicited and unsolicited manuscripts focusing on food chemistry are considered for publication in this series. Submitted manuscripts will be reviewed and decided by the series editor, Prof. Dr. Salvatore Parisi.

To submit a proposal or request further information, please contact Dr. Sofia Costa, Publishing Editor, via sofia.costa@springer.com or Prof. Dr. Salvatore Parisi, Book Series Editor, via drparisi@inwind.it or drsalparisi5@gmail.com

More information about this subseries at http://www.springer.com/series/11853

Teresa De Pilli · Antonietta Baiano ·
Giuseppe Lopriore · Carlo Russo ·
Giulio Mario Cappelletti

Sustainable Innovations in Food Packaging

 Springer

Teresa De Pilli
Department of Agriculture, Food, Natural
Resources and Engineering (DAFNE)
University of Foggia
Foggia, Italy

Antonietta Baiano
Department of Agriculture, Food, Natural
Resources and Engineering (DAFNE)
University of Foggia
Foggia, Italy

Giuseppe Lopriore
Department of Agriculture, Food, Natural
Resources and Engineering (DAFNE)
University of Foggia
Foggia, Italy

Carlo Russo
Department of Economics, Management
and Territory (DEMeT)
University of Foggia
Foggia, Italy

Giulio Mario Cappelletti
Department of Economics, Management
and Territory (DEMeT)
University of Foggia
Foggia, Italy

ISSN 2191-5407   ISSN 2191-5415   (electronic)
SpringerBriefs in Molecular Science
ISSN 2199-689X   ISSN 2199-7209   (electronic)
Chemistry of Foods
ISBN 978-3-030-80935-5   ISBN 978-3-030-80936-2   (eBook)
https://doi.org/10.1007/978-3-030-80936-2

This Springer imprint is published by the registered company Springer Nature Switzerland AG
The registered company address is: Gewerbestrasse 11, 6330 Cham, Switzerland

Preface

This brief concerns the interesting sector of sustainable food packaging. The ambitious goal to develop a sustainable production system involves the reduction of emissions, the efficient use of resources, and the transition to renewable energy. The bioeconomy theorizes a model that would like to reduce impacts and risks associated with the use of non-renewable resources considering the life cycle of products. The European Union furthers packaging from renewable sources focused on bio-based materials. Packaging is an important key to satisfy the increasingly pressing and urgent request of sustainable food production and consumption on the industrialized countries to reduce a minimum the environmental impact of packaged food. A new sustainable packaging should guarantee the reuse of whole waste material and the loss of food safety and quality during storage by preventing food-borne diseases and food chemical contamination. Also, it must consider the dramatic problem correlated with persistent plastic waste accumulation as well as the saving of oil and food material resources. This book presents the main innovations of food packaging that aim to win the pressing international challenges related to food and plastic waste reduction and end-of-life issues of persistent materials. Among potential solutions, the production of microbial biodegradable polymers and the use of by-products and waste of agricultural and food industries seem a promising route to create an innovative and productive waste-based food packaging economy, separating the food packaging industry from fossil stocks and allowing biopolymers to return to the soil. Moreover, to understand the economical reliability of these innovations, different analyses—life cycle assessment or LCA, life cycle costing or LCC, and externality assessment or ExA—are discussed to assess the impacts along the whole chain by means of an integrated approach.

Chapter 1 has been realized by Prof. Antonietta Baiano, while Chap. 2 has been written by Prof. Giuseppe Lopriore, Prof. Carlo Russo, and Prof. Giulio Mario Cappelletti. Finally, Chaps. 3 and 4 have been realized by Prof. Teresa De Pilli.

Foggia, Italy

Teresa De Pilli
Antonietta Baiano
Giuseppe Lopriore
Carlo Russo
Giulio Mario Cappelletti

Contents

Chapter 1
An Overview on the Environmental Impact of Food Packaging

Abstract Today, a wide range of traditional materials are used for food packaging applications and new packaging materials are constantly being developed. Although food packaging materials exert positive effects on the environment by preventing spoilage and reducing food waste, after use, their disposal can remarkably affect the environment especially if the 3Rs approach is not followed. The aim of this Chapter is to give an insight on the environmental impacts of packaging materials intended for food use throughout their life cycle. The highest contributions to the packaging waste were supplied by paper/paperboard (50.8%) that also shows the highest recycling rate together with plastics (about 73%). The Regulatory Frameworks establish minimum recycling targets for various packaging materials. The packaging environmental impacts, evaluated through the application of quali-quantitative methods such as LCA, eco-design and carbon footprint, must be accompanied by the evaluation of the impacts of the food packaging system. Some new packaging solutions contribute to various impact categories more than the conventional ones but, being able to considerably prevent food losses, they minimize the environmental impact of the contained food.

Keywords Circular economy · Disposal · Environmental sustainability · LCA · Packaging waste · Recycle · Reuse

Abbreviations

CCME	Canadian Council of Ministers for the Environment
CO_2	Carbon dioxide
COD	Chemical oxygen demand
EEA	European Economic Area
COMPASS	Comparative Packaging Assessment
EFTA	European Free Trade Association
EU	European Union
GHG	Greenhouse Gas
ISO	International Organization for Standardization
LCA	Life Cycle Assessment
LCI	Life Cycle Inventory

© The Author(s), under exclusive license to Springer Nature Switzerland AG 2021
T. De Pilli et al., *Sustainable Innovations in Food Packaging*, Chemistry of Foods,
https://doi.org/10.1007/978-3-030-80936-2_1

LCIA	Life Cycle Impact Assessment
LLDPE	Linear Low-Density Polyethylene
LDPE	Low-Density Polyethylene
MSW	Municipal Solid Waste
PtP	Package-to-Product
PIQET	Packaging Impact Quick Evaluation Tool
PA	Poly-Amide
PAH	Polycyclic Aromatic Hydrocarbons
PE	Polyethylene
PE-g-MA	Polyethylene-grafted maleic anhydride
PET	Polyethylene terephthalate
PHA	Polyhydroxyalkanoate
PP	Polypropylene
3Rs	Reduce, Reuse and Recycle
R&D	Research and development
SBS	Solid Bleached Sulfate
SO_2	Sulphur Dioxide
SAP	System Analysis Program Development
TPS	Thermoplastic starch
TiO_2	Titanium dioxide
TFEU	Treaty on the Functioning of the European Union
UV	Ultraviolet
U.S.	United States

1.1   Food Packaging and Environmental Impact Generation

Packaging is one of the most important operations within the food manufacturing process and packaging is also the generic name of the object intended to contain food. The main functions of food packaging are: contain the product; protect it from biological, chemical and physical injuries as well as from outside influences and damages that can affect its safety and quality and reduce its shelf life; facilitate food handling and contribute to efficient distribution, sales and consumption; host the product label and, possibly, tamper indicators (Marsh and Bugusu 2007). A packaging system can consist of three main parts known as primary, secondary and tertiary packaging. The primary packaging put into direct contact with the product, borders the selling unit. The secondary or collection packaging is used to protect the primary packaging and to carry quantities of primary packaged goods. The tertiary packaging is used for bulk handling in warehousing and transportation. Further extra functions of food primary packaging can be found in the categories of active and intelligent packaging systems. The first materials are able to absorb or release substances to improve safety/quality of packaged foods or to extend their shelf life, while intelligent packaging is able to both monitor and give information about the condition of the packaged foods during transport and storage.

Nowadays, a wide range of materials are commonly used for food packaging applications and new packaging materials are constantly being developed. The conventional materials for food packaging include (Geueke et al. 2018): metals such as aluminium and steel, in the form of cans, tubes, containers, films, caps and closures; paper and paperboards, commonly used in corrugated boxes, milk cartons, folding cartons, paper plates and cups, bags and sacks and wrapping paper; glass as narrow-neck bottles and wide-opening jars and pots; plastics such as polyolefins (PE and PP), polyester (for example, PET), polyvinyl chloride, polyvinylidene chloride, polystyrene, PA and ethylene vinyl alcohol. These plastics have become the most common food packaging materials and are currently used to produce bottles, trays, bags, foils, cups, pots, pouches and bowls (Marsh and Bugusu 2007). Innovations in food packaging materials concern: the use of nanoparticles, to improve the mechanical and thermal properties of packages and monitor the safety and quality of the foods; the development of biodegradable polymers, considered as alternatives to the traditional plastics for more sustainable packaging solutions. Biodegradable polymers can be categorized in natural (based on starch, protein, chitosan, cellulose), synthetic (PHA, bacterial cellulose), and non-natural (polyvinyl alcohol, polybutylene adipate-co-terephthalate and polylactide, as examples) (Rydz et al. 2018).

Food packaging exerts undoubted positive effects on the environment by reducing food waste throughout the supply chain. Nevertheless, it is impossible to ignore that, in the developed countries, food packaging represents approximately two-thirds of all packaging (Shin and Selke 2014) and, although there are packaging design options able to decrease both food waste and the environmental impact of the packaging itself, most packaging systems have noticeable environmental impacts, especially if designed for single use. The environmental negative impacts of packaging systems start with their creation, due to the consumption of resources such as energy, water and raw materials necessary for their manufacture and the release of greenhouse gases (GHG), heavy metals, toxic contaminants, wastewater and sludge. Further impacts are generated at the packaging end-of-life, as a consequence of their disposal. There is great variability of the environmental impact ratio between packaging and food for different products. Based on a survey of the food LCA literature, Heller et al. (2018) calculated the food-to-packaging ratio in terms of GHG emissions for several products aggregated by food type (beverages, cereals and grains, dairy, fish and seafood, fruit, legumes and nuts, meat, vegetables). This ratio ranges from 0.06 (wine) to 780 (beef) and cereals, dairy. Fish, seafood and meats show the highest values of this ratio. In presence of low values of the environmental impact ratio, it is preferable to focus attention on reducing the impact of the packaging, since the food waste reduction would not have significant influence on the total system environmental performance. Instead, at high ratio values, the food waste reduction would give larger system benefits.

1.2   Statistics on Food Packaging Waste

The food packaging market is the largest segment of the packaging industry, accounting for 282.6 billion U.S. dollars in 2016 and with further growth expectations due to the increase in the consumption of convenience foods. The growth of individual food segments exerts a remarkable influence on the type and volume of food packaging. Currently, the driven features of the food packaging market include easy opening, portability and easy disposability while visual appeal and convenience are additional aspects. The global weight of municipal solid waste (MSW) was 2.01 billion tons in 2016 and is expected to increase up to 3.40 billion tons in 2050. About 30–35% of the MSW are represented by packaging waste and almost 60% of packaging is for food application (Nemat et al. 2020). According to the United States Environmental Protection Agency (EPA 2019), the total amount of food and non-food packaging and containers in MSW was around 80,000 thousand U.S. tons in 2017 (Table 1.1). The rates of recycling, combustion with energy recovery and landfilling were 50.1, 9.8 and 40.1%, respectively. The highest contributions (by weight) to the packaging waste were supplied by paper and paperboard (50.8%) and plastic (18%). The materials showing the highest rate of recycling were paper/paperboard (73.3%) and steel (73.1%) while glass was surprisingly recycled by only 33.9%. As well as for miscellaneous (80%), the highest rates of landfilling were observed for plastic (70%), wood (59.2%), aluminium (53.4%), and glass (53.1%). Energy was recovered by combustion of miscellaneous materials, plastic, wood, aluminium and glass in percentages ranging from 13 to 20%. In EU, in the same reference year, 77.5 million tons of packaging waste were generated and the most represented materials were paper/cardboard (41%), plastic (19%), glass (18%), wood (17%), and metal (5%). The total generated packaging waste rose by 9.3% from 2007 to 2017. The recovery (including incineration with energy recovery) and recycling rates were 81.7 and 67.5%, respectively. From 2007 to 2017, the recovery and recycling rates rose by 7.9 and 8.4%, respectively. In 2017, within EEA/EFTA countries, Finland and Belgium held the highest recovery (112.1%) and recycling (83.8%) rate, respectively. Malta had the lowest recovery and recycling rates (35.6%) (Eurostat 2020).

1.3   Strategies to Reduce Packaging Environmental Impacts

In order to reduce the environmental impacts of food packaging, the preliminary action consists in elimination of the unnecessary packaging, i.e. that whose absence does not affect food safety and quality. In addition, the 3Rs hierarchy (Fig. 1.1) has become a significant policy approach to achieve sustainable consumption and production, (Ahmadi 2017). The first R stays for 'Reduce' and, as can be inferred from the figure, is the most important of the 3Rs. It represents a preventive action

Table 1.1 Containers/packaging in MSW (EPA 2019)

Management system	Total	Glass	Steel	Aluminium	Paper and paperboard	Plastic	Wood	Miscellaneous[a]
Generation	80,080	8930	2010	1890	41,060	14,490	11,350	350
Recycling	40,090	3030	1470	620	30,080	1890	3000	–
Composting	–	–	–	–	–	–	–	–
Combustion with energy recovery	7860	1160	100	260	2160	2470	1640	70
Landfilling	32,130	4740	440	1010	8820	10,130	6710	280

Data are referred to 2017 and are expressed as thousands of U.S. tons
[a]Bags made of textiles; leather

aimed to both improve manufacturing methods thus reducing packaging and influence consumers' behaviour so that they demand more products and less packaging (European Commission 2010). Reduction occurs by: minimizing the weight and volume of materials used; employing bulk delivery systems and bulk retail sales; using fewer toxic chemicals in the packaging production; using alternative material selection; using multi-layered and multi-material packaging systems, although they are non-recyclable. According to the second R (also in order of importance), packaging should be: reusable either for the same purpose or for a new one, refillable, returnable and durable. Reuse is preferable to recycle (the third R in importance) since the first does not require further reprocessing (Ahmadi 2017). Packaging materials are considered recyclable if they can be collected, segregated, processed (into its original packaging type or into another recyclable/non-recyclable product), and marketed. Based on the recycling effectiveness, it is possible to distinguish upcycling, recycling and downcycling. Through upcycling, a material is converted into an object of greater value than the original ones. Recycling, instead, puts a material back into the same life cycle, meaning that the true recycling gives origin to a product similar to the original ones. Downcycling is defined as the recycling of the material into a lower quality matter, which is in turn used to give rise to a lower-grade product. It is often determined by the nature of the material that prevents it from maintaining its native characteristics and durability after reprocessing. Other ways to reduce food packaging waste include: composting, i.e. the aerobic or biological degradation of organic material; combustion, useful to obtain heat or electricity from the burning of packaging that cannot be recycled or composted; landfilling of what remains after recycling and combustion.

Materials such as paper and board, metals and glass can be conveniently recycled but with substantial differences among materials. Concerning paper, 3.5 recycles are generally applied in Europe, although up to 7 cycles are technically feasible. The fractionation of the collected material into food grade and non-food

Fig. 1.1 The 3R's packaging hierarchy

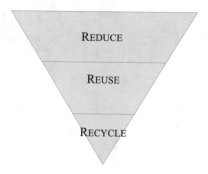

REDUCE

REUSE

RECYCLE

grade streams before recycling is not common for waste paper. Furthermore, thousands of chemical compounds such as additives, printing inks, adhesives, photoinitiators, solvents, plasticisers and pigments are present in waste paper and can be retrieved into the recycled product (Pivnenko et al. 2015; Van Bossuyt et al. 2016). Other contaminations can occur during waste management (Geueke et al. 2018). Based on these findings, the recycled paper and board should not be put in direct contact with food.

Concerning metal containers, cans are commonly recycled. Aluminium can be almost indefinitely recycled although repeated recycling can result in accumulation of unwanted metals and metalloids, alloying elements, or coating residues (Løvik and Müller 2014). The average global recycling rates are around 70%.

An efficient recycling of glass requests that containers are separated by colour and origin. Then, glass must be mixed with virgin raw materials, melted and formed into new products.

Thermoplastic polymers are generally submitted to a mechanical recycling although they can be also recycled by chemical processes. During mechanical recycling, heating can cause the breakage of intramolecular bonds thus resulting in a lower molecular weight distribution and, consequently, in changes of both mechanical and optical properties of the recycled polymers (Ignatyev et al. 2014), without counting the alterations that polymers can undergo due to the contact with acids, high temperatures and UV irradiation occurring during use and waste managements (Geueke et al. 2018). The careful separation of plastic polymers by type of materials is a prerequisite for an effective recycling (Hopewell et al. 2009). Since plastic waste can be contaminated by degradation products of polymers and additives, substances deriving from previous use and from the environment and cross-contaminations from waste disposal, there is the need for cleaning steps performed through treatments such as high temperatures, vacuum/inert gas, or extraction with supercritical carbon dioxide (Welle 2011). Thermoset polymers generally cannot be recycled since they form permanent chemical bonds and cross links and maintain strength and shape even if submitted to prolonged heating. However, new recyclable thermoset materials have been developed, including epoxy resins that, under controlled heating, form crosslinked networks that can be repeatedly reshaping and polyurethanes, which have transient properties that allow their reprocessing or recycling (Wang et al. 2020).

Packaging consisting of several layers of plastics (plastic multilayers) or of different materials including paperboard, aluminium and plastic (multimaterial multilayers and laminated packaging) are largely used to contain food and beverages since the various layers give the container-specific requirements of permeability, resistance, mechanics, printability, etc. Beverage cartons are examples of multilayer material and can be recycled by separating paperboard, from plastic (usually PE) and aluminium. Carton fibre can be mixed with virgin material to produce new paper products (also food packaging not in direct contact with food) while plastic can be used for energy recovery and aluminium can be recycled (Geueke et al. 2018).

1.4 Methods for the Assessment of the Environmental Impact of Packaging

The life cycle of packaging includes 4 steps namely design, manufacture, usage and disposal. With the exception of the first one, each of these steps exerts more or less remarkable environmental impacts and, for this reason, the environmental impact assessment should be included in the design step. First of all, it is necessary to create a generic framework with all the requirements for packaging. Then it is necessary to establish the most appropriate parameters for the evaluation of the product packaging environmental. A list of 27 parameters, divided into the following life cycle steps, were established (Singh et al. 2014): preproduction (3 parameters), production (9), usage (5), and end-of-life (10). In more recent work, weights were assigned to all the parameters and those having values lower than a certain cutoff weight were discarded (Singh et al. 2018). In this way, the original 27 parameters were reduced to 16, five of them, namely 'the distance between major resources and production unit', 'the percentage of recycling material used', 'average life', 'the energy required for recovering of material', and 'the energy required for recyclability' were not considered by the already-existing tools for the packaging environmental impact assessment used as benchmarks: COMPASS; Walmart's packaging guidelines; PIQET; CCME guidelines; SAP Packing Instructions; West Rock sustainability approach; Nextek Limited sustainable solutions for polymer recycling; 360 environmental limited packaging obligation calculator; Valpak's LCA-based approach to sustainability. The selected parameters, categorized on the basis of their priorities in High (5 parameters), Moderate (8), and Low (3), indicated that the evaluation of the 'energy required for manufacturing', the 'material used in manufacture', the 'depletion of resources in the extraction of materials', the 'depletion of fresh water in the manufacture', and the 'pollutants released during manufacture' are of fundamental importance.

Two types of approaches are generally used to estimate the environmental effects of packaging materials. The quantitative approach relies on statistics concerning a great amount of data and includes: monetary methods (for example, cost and benefit analysis) to derive the cost associated with impacts; estimation of pollution emission and resource and energy consumption (for example, material flow analysis and LCA). The qualitative approach is based on case studies, interviews and reports, administration of questionnaires and creation of focus groups. Quantitative methods are generally preferred in decision making since they help to evaluate better relationships between cause and effect while qualitative methods are able to reveal the subject perception of environmental impacts. A combined quali-quantitative approach supplies a multidimensional environmental evaluation framework which is a more effectively guide (Huang and Ma 2004). Including the already cited ones, a number of tools are available to assess the environmental impacts of packaging. They can be classified into: conventional LCA; simplified (or streamlined) LCA, usable without extensive training; and Non-LCA software tools, which do not follow the life cycle approach (Pauer et al. 2017). Conventional LCA is increasingly used both in evaluating the potential environmental impacts of any product throughout its life cycle and in decision making. The general guidance to perform an LCA study is stated by ISO 14040:2018 and ISO 14044:2018. The LCA approach includes the following 4 steps: definition of goal and scope of the study; LCI, to collect data and quantify the relevant inputs and outputs of the product system; LCIA, aimed to classify data into categories of impacts, using characterization factors and according to the area of interest, namely human health, natural resources and natural environment; interpretation of results. A number of software packages have been developed to facilitate assembling of all the information and doing the required calculations contemplated by LCA studies. Some of these (for example, SimaPro and GaBi) are full-fledged ISO 14040/14044-compliant programs while others, such as COMPASS and Package Modelling, are simpler but show some limits. Moreover, by comparing the various software systems using a common set of packaging materials, results from various LCA software systems are often in disagreement on the environmental impact of containers so that the software choice can influence the packaging-system decision making (Speck et al. 2015).

Eco-design is the inclusion of the environmental aspects of a product into the design process of the same product, in order to improve its environmental performance throughout the entire life cycle and to comply with legal requirements (Directive 2009/125). Eco-design strategies concern material selection, optimization of volumes to decrease the transportation impact), and multifunctionality of the packaging to increase its duration and to attract the customer. An eco-design process, as defined by ISO 14006:2020, includes the following 6 steps: 'specify product functions'; 'environmental assessment of products' generally through LCA; 'strategies of improvement'; 'environmental objectives'; 'product specification'; 'technical solutions' (Navajas et al. 2017).

Carbon Footprint is another method suitable to both investigate the environmental impact of packaging and communicate its environmental performance in terms of GHG. Although LCA methodology is more complete since it considers several impact categories, 'Carbon Footprint' is more understandable by customers because climate change and global warming issues as well as CO_2 units are generally known (ISO 14067:2018; Sanyé-Mengual et al. 2014). A Package-to-Product (PtP) indicator based on Carbon Footprint and LCA has been recently developed and tested to assess the environmental impacts of the packaging together with the corresponding product throughout their entire life cycle (Šerešová and Kočí 2020).

1.5 Environmental Impacts Caused by Food Packaging

The environmental impacts of a food packaging material start with its design and ends with its disposal. First of all, the manufacture of packaging causes the consumption of raw materials and semi-finished products such as petroleum and derivatives, chemicals, minerals, wood, as well as the use of energy and water.

Moreover, the packaging production process causes the release into the air of gases (GHG, sulfur oxides and nitrogen oxides), fine particulate, PAH, volatile organic compounds and heavy metals such as arsenic and lead, as well as the generation of wastewater and/or sludge that can contain polluting substances. After their use, packaging materials are discarded and, if wastes are not properly treated, they can cause air, water and soil pollution, with remarkable negative effects on humans, fauna and flora. The same landfills release ammonia and hydrogen sulfide while incineration causes the emission of heavy metal, hydrogen chloride, sulfur dioxides, nitrous oxides and particulates.

The following paragraphs give an overview of some recent researches performed on the environmental impacts of packaging traditionally used to contained various categories of foods and newly developed packaging materials.

1.5.1 Environmental Impact of Traditional Food Packaging

Bertoluci et al. (2014) applied LCA to evaluate the environmental impacts of three olive packaging solutions, namely doypacks, glass jars and steel cans, in five EU countries (France, Germany, Italy, Spain and Sweden). The first result was that the environmental performance of each type of packaging can considerably change from one country to another due to different household waste collection rates and technologies for waste treatment (recycling/incineration). However, glass jar was the worst packaging solution in all the considered countries while doypacks showed the lowest environmental impacts, although they are made of a non-renewable and non-recyclable multilayer material. The researchers also performed a qualitative analysis of user expectations, demonstrating that the environmentally better solution did not satisfy the consumer preferences that were addressed to glass for several reasons: a glass package can be used to store the product once opened; the product can be easily extracted; containers are generally associated with aspects such as quality and security.

The LCA assessment applied to reusable plastic and glass food savers in Europe highlighted that glass containers have 12–64% higher impacts than the plastic ones and their lifespan should be 3.5 times greater than the actual value, having the same environmental footprint of the plastic containers. Since the use step contributes for more than 40% to the whole impacts as a consequence of washings, it is possible to obtain an impact decrease of 12–27% by improved hand dishwashing, for example lowering the amount of water, energy and detergents used (Gallego-Schmid et al. 2018).

The work of Lewis and coworkers applied LCA to compare 2 types of frozen food packages: the common PET tray-and-film package, placed inside a SBS paperboard box; and the Traytite® package, composed of SBS paperboard coated with a PET film (Lewis et al. 2008). The results demonstrate that Traytite®: consumes less raw materials, water, chemicals and total energy than tray-and-film package; emits less SO_2 and methane in the air; has a lower COD; has lower total climate change potential. Furthermore, the greatest percentage of climate change impact for both packaging materials is due to the paper pulping process. Finally, the only packaging component that can be significantly recycled is the paperboard carton of the tray-and-film system. Traytite® trays and PET components of tray-and-film systems are generally not recycled.

1.5.2 Environmental Impact of Innovative Food Packaging

Nano-packaging is considered as a promising technology for food preservation since the incorporation of nanomaterials into polymers can improve their barrier properties or, as in case of metal nanoparticles, can exert antimicrobial activities with the potential to reduce food waste. Nevertheless, the production of nanomaterials requires extra input of energy, water and other resources, generating further emissions and wastes.

Zhang and coworkers performed a LCA study to evaluate the carbon footprint of 4 types of nano-packaging materials considering the potential food waste reduction due to their ability to extend shelf life (Zhang et al. 2019): LLDPE + PE-g-MA + nanoclay; LLDPE + PET + nanoclay; LDPE + 95% TiO_2 + 5% nano-silver; PA 6 + LDPE + nanoclay. The total carbon footprint of different packaging systems was determined considering the carbon footprints of nanomaterials (positive) and the reduced carbon footprint of food saved from becoming waste (negative). It was negative for all the considered packaging systems, indicating that the food waste saved by the use of nano-packaging was higher than the environmental impact generated by the incorporated nanomaterials during their life cycle.

PHA-TPS is a new multilayer material that shows interesting oxygen barrier properties in addition to high biodegradation rates in landfill or composting and ability to biodegrade in sea water. PHA is produced through bacterial fermentation of a carbon source under limited nutrient conditions that determine the intracellular accumulation of the polymer. Dilkes-Hoffman et al. (2018) compared the global warming potential of two types of packaging materials, PHA-TPS and PP, in prolonging the shelf life of beef and cheese. They found that, although PHA-TPS packaging is produced from materials that absorb CO_2 during growth, landfilling of this new material causes a higher global warming potential than PP due to methane emissions. As a consequence, landfill methane capture efficiency is an important aspect when biodegradable packaging is used (instead non-biodegradable packaging is inert). When landfill methane is not captured, the GHG emissions of the PHA-TPS packaging are 7% higher than those of PP packaging if food wastage is considered to be equal for the two types of containers. Instead, it is sufficient that the PHA-TPS package is able to extend food shelf life thus reducing food waste by 6% to counterbalance the emissions associated with the biodegradable package.

Zhang and coworkers applied LCA to evaluate the environmental impact of a food packaging system made of beef and a thymol/carvacrol active packaging in comparison with the conventional one (Zhang et al. 2015). Regarding the impact categories of 'acidification potential' and 'eutrophication potential', it is sufficient that the active packaging prevents the beef loss of only 0.1% to make it convenient from an environmental point of view. Concerning the impact categories represented by 'energy demand', it is necessary that the active packaging reduce beef losses of 0.6%.

References

Ahmadi M (2017) Evaluating the performance of 3Rs waste practices: case study-region one municipality of Tehran. Adv Recycl Waste Manag 2:2. https://doi.org/10.4172/2475-7675. 1000130

Bertoluci J, Leroy Y, Olsson A (2014) Exploring the environmental impact soft olive packaging solutions for the European food market. J Clean Prod 64:234–243. https://doi.org/10.1016/j. jclepro.2013.09.029

Dilkes-Hoffman LS, Lane JL, Grant T, Pratt S, Lant PA, Laycock B (2018) Environmental impact of biodegradable food packaging when considering food waste. J Clean Prod 180:325–334. https://doi.org/10.1016/j.jclepro.2018.01.169

Directive 2009/125 (2009) Establishing a framework for the setting of ecodesign requirements for energy-related products (recast). Official J Eur Union L285:10–35

EPA (2019) Facts and figures about materials, waste and recycling. Available https://www.epa. gov/facts-and-figures-about-materials-waste-and-recycling/containers-and-packaging-product-specific-data. Accessed 19 Aug 2020

European Commission (2010) Being wise with waste: the EU's approach to waste management. Available https://doi.org/10.2779/93543. Accessed 30 Sept 2020

Eurostat (2020) Packaging waste statistics. Available https://ec.europa.eu/eurostat/statistics-explained/index.php/Packaging_waste_statistics#Time_series_of_packaging_waste_generation_and_treatment. Accessed 19 Aug 2020

Gallego-Schmid A, Mendoza JMF, Azapagic A (2018) Improving the environmental sustainability of reusable food containers in Europe. Sci Total Environ 628–629:979–989. https://doi.org/10. 1016/j.scitotenv.2018.02.128

Geueke B, Groh K, Muncke J (2018) Food packaging in the circular economy: overview of chemical safety aspects for commonly used materials. J Clean Prod 193:491–505. https://doi. org/10.1016/j.jclepro.2018.05.005

Heller MC, Selke SEM, Keoleian GA (2018) Mapping the influence of food waste in food packaging environmental performance assessments. J Ind Ecol 23(2):480–495. https://doi.org/ 10.1111/jiec.12743

Hopewell J, Dvorak R, Kosior E (2009) Plastics recycling: challenges and opportunities. Philos Trans R Soc Lond B 364(1526):2115–2126. https://doi.org/10.1098/rstb.2008.0311

Huang CC, Ma HW (2004) A multidimensional environmental evaluation of packaging materials. Sci Total Environ 324:161–172. https://doi.org/10.1016/j.scitotenv.2003.10.039

Ignatyev IA, Thielemans W, Vander Beke B (2014) Recycling of polymer: a review. Chemsuschem 7:1579–1593. https://doi.org/10.1002/cssc.201300898

ISO 14006 (2020) Environmental management systems—guidelines for incorporating ecodesign. International Organization for Standardization, Geneva

ISO 14040 (2018) Environmental management—life cycle assessment—principles and framework. International Organization for Standardization, Geneva

ISO 14044 (2018) Environmental management—life cycle assessment—requirements and guidelines. International Organization for Standardization, Geneva

ISO 14067 (2018) Greenhouse gases—carbon footprint of products—requirements and guidelines for quantification. International Organization for Standardization, Geneva

Lewis A, Keaton L, Scoville D, Espenschied T, Griffin L, Weis A, Li K (2008) Comparative life cycle assessment of Traytite® and tray- and film frozen food packaging. Available https://ie. unc.edu/files/2016/03/frozen_food_final_report.pdf. Accessed 21 Aug 2020

Løvik AN, Müller DB (2014) A material flow model for impurity accumulation in beverage can recycling systems. In: Grandfield J (ed) Light metals. Springer, Switzerland, Cham, pp 907–911

Marsh K, Bugusu B (2007) Food packaging—roles, materials, and environmental issues. J Food Sci 72(3):R39–R55. https://doi.org/10.1111/j.1750-3841.2007.00301.x

Navajas A, Uriarte L, Gandía LM (2017) Application of eco-design and life cycle assessment standards for environmental impact reduction of an industrial product. Sustainability 9:1724. https://doi.org/10.3390/su9101724

Nemat B, Razzaghi M, Bolton K, Rousta K (2020) The potential of food packaging attributes to influence consumers' decisions to sort waste. Sustainability 12:2234. https://doi.org/10.3390/su12062234

Pauer E, Heinrich V, Tacker M (2017) Methods for the assessment of environmental sustainability of packaging: a review. IJRDO—J Agric Res 3(6):33–62. Available http://www.ijrdo.org/index.php/ar/article/view/86. Accessed 21 Aug 2020

Pivnenko K, Eriksson E, Astrup TF (2015) Waste paper for recycling: overview and identification of potentially critical substances. Waste Manag 45:134–142. https://doi.org/10.1016/j.wasman.2015.02.028

Rydz J, Musioł M, Zawidlak-Węgrzyńska B, Sikorska W (2018). Present and future of biodegradable polymers for food packaging applications. In: Grumezescu AM, Holban AM (eds) Biopolymers for food design. Academic Press, Massachusetts, pp 431–467. https://doi.org/10.1016/B978-0-12-811449-0.00014-1

Sanyé-Menguala E, García Lozanoa R, Oliver-Solàa J, Gasola CM, Rieradevall J (2014) Ecodesign and product carbon footprint use in the packaging sector. In: Muthu SS (ed) Assessment of carbon footprint in different industrial sectors, vol 1. Springer, Switzerland, pp 221–245

Šerešová M, Kočí V (2020) Proposal of package-to-product indicator for carbon footprint assessment with focus on the Czech Republic. Sustainability 12:3034. https://doi.org/10.3390/su12073034

Shin J, Selke SEM (2014) Food packaging. In: Clark S, Jung S, Lamsal B (eds) Food processing: principles and applications. Wiley, New Jersey, pp 249–273. https://doi.org/10.1002/9781118846315.ch11

Singh S, Kumar J, Rao PVM (2014) Identification of parameters for environmental impact assessment of product packaging. In: KES transactions on sustainable design and manufacturing i sustainable design and manufacturing, pp 419–431

Singh S, Kumar J, Rao PVM (2018) Parameters for environmental impact assessment of product packaging: a Delphi study. J Packag Technol Res 2:3–15. https://doi.org/10.1007/s41783-018-0027-4

Speck R, Selke S, Auras R, Fitzsimmons J (2015) Choice of life cycle assessment software can impact packaging system decisions. Packag Technol Sci 28:579–588. https://doi.org/10.1002/pts.2123

Van Bossuyt M, Van Hoeck E, Vanhaecke T, Rogiers V, Mertens B (2016) Printed paper and board food contact materials as a potential source of food contamination. Regul Toxicol Pharmacol 81:10–19. https://doi.org/10.1016/j.yrtph.2016.06.025

Wang S, Yang Y, Ying H, Jing X, Wang B, Zhang Y, Cheng J (2020) Recyclable, self-healable, and highly malleable poly(urethane-urea)s with improved thermal and mechanical performances. ACS Appl Mater Interf 12(31):35403–35414. https://doi.org/10.1021/acsami.0c07553

Welle F (2011) Twenty years of PET bottle to bottle recycling—an overview. Resour Conserv Recycl 55(11):865–875

Zhang BY, Hortal M, Dobon A, Bermudez JM, Lara-Lledo M (2015) The effect of active packaging on minimizing food losses: Life Cycle Assessment (LCA) of essential oil component-enabled packaging for fresh beef. Packag Technol Sci 28:761–774. https://doi.org/10.1002/pts.2135

Zhang BY, Tong Y, Singh S, Cai H, Huang JY (2019) Assessment of carbon footprint of nano-packaging considering potential T food waste reduction due to shelf life extension. Resour Conserv Recy 149:322–331. https://doi.org/10.1016/j.resconrec.2019.05.030

Chapter 2
Environmental and Socio-Economic Sustainability of Packaging from Agricultural By-Products

Abstract According to the circular economy and sustainable development concepts, this chapter aims to analyze the sustainability issues about agro-food by-products valorisation in order to obtain bio-based compounds as material for food packaging. The overview focuses on experimentations in which bio-polymers were processed and tested with the purpose of evaluating the characteristics and properties in terms of food preservation and active functions. Despite a lot of data about technical features were provided from these studies, in many cases, information on sustainable aspects were missed. This represents a crucial issue because any solution to be really sustainable needs to be assessed in an overall perspective with a life cycle approach. However, in the evaluation of the sustainability of bio-based compounds coming from the valorisation of agro-food by-products it is fundamental to consider elements such as availability in time and space, transports, impacts of processes, as well as, hidden costs along the chain and impact on local communities. At the same way, indirect effects of the packaging, such as the capacity of reducing food waste, should be also taken into account. In the chapter, a simulation about the possible way to assess sustainability of bio-based material was purposed by using life cycle assessment with a consequential approach. First of all, due to the fact that agro-food by-products derive from the same system of production of main agro-food product, the importance of setting an allocation procedure was highlighted from the study as a crucial issue before calculating impacts. Successively, the sustainable performances are accounted by including the avoided impacts related to the production of petroleum-based packaging material.

Keywords Allocation factors · By-product allocation procedure · GWP · Life cycle assessment · Packaging additive · Packaging main ingredient · Packaging secondary ingredient

Abbreviations

CO_2	Carbon dioxide
CNF	Cellulose nanofibrils
DDGS	Distiller-dried grains with solubles
GWP	Global Warming Potential

GHG   Greenhouse gases
HACCP   Hazard Analysis and Critical Control Points
HDPE   High density polyethylene
KNN   k-Nearest Neighbour
LCA   Life Cycle Assessment
LDPE   Low density polyethylene
MAE   Microwave assisted extraction
OOMW   Olive oil mill wastewater
PHA   Polyhydroxyalkanoates
PHBV   Polyhydroxybutyrate-co-valerate
PLA   Polylactic acid
PVA   Polyvinyl alcohol
PE   Polyethylene
PP   Polypropylene
TRL   Technology Readiness Levels

2.1 Achieving Sustainability in Food Packaging

In order to achieve sustainability in food packaging, it is important to consider the overall systems of food and packaging production. Often, only technical aspects and function performances are considered when studying packaging solutions (Valdes et al. 2014).

According to the new paradigm of sustainable development and circularity of the economy, a complete sustainable analysis should include not only direct implications (such as eco-design or impact of transports) but also indirect effects (such as the role that packaging could have in dealing with important issues like food losses due to the shelf-life extension). Thus, it could entail changing behaviours and consumers habits with crucial consequences able to guide society towards the achieving of real sustainable goals (Brennan et al. 2020; Chisenga et al. 2020; Conte et al. 2015; Zhang et al. 2015).

In assessing sustainability of packaging solutions coming from agro-food by-products, it is worth noting how the re-use as secondary raw material could improve the performances of the sustainability of the main product. This, from a life cycle perspective, means that just as the by-product leaves the system boundaries of the main product, so does its impacts could be eliminated by the sustainability assessment of the main product (EC 2018; Ekvall et al. 2016; JRC-IES 2010; Payen et al. 2018).

At the same way, for considering really sustainable a packaging material obtained by an agro-food by-product the availability of this latter and the role of transportation in the overall process should be well investigated, as well as the management of its disposal (Varžinskas and Markevičiūtė 2020).

Starting from these premises this chapter aims to investigate how sustainability is dealt with in reusing agro-food by-products as packaging material. After a brief review about papers which face the issue of assessing sustainability to innovative bio-based packaging solutions, the study purposes an example of application of sustainability assessment by using life cycle assessment methodology with a consequential approach (Table 2.1).

Table 2.1 Agro-food by-products used as inherently biodegradable packaging materials

References	By-products
Inherently biodegradable agro bio-based for packaging materials	
Ren et al. (2021)	Sweet potato starch
Lauer and Smith (2020)	Starch
Hilliou et al. (2016)	Cheese whey and olive oil mill wastewater
Wu et al. (2019)	Pomelo peel flour and tea polyphenol
Sá et al. (2020)	Cashew apple and cashew tree pruning fibre
Souza Filho et al. (2020)	Wheat bran and straw
Jafarzadeh et al. (2020)	Avocado seed and peel, cacao bean husk, coconut palm husks, mango kernel, potato peel
Adilah et al. (2018)	Mango peels extract and fish gelatine
Andres et al. (2017)	Tomato pomace
da Rosa et al. (2019)	Olive leaf extract added to Carrageenan film
Nerantzis and Tataridis (2006)	Vine prunings and grape pomace
Sogut and Seydim (2018)	Grape seed extract
Cerón Gómez (2020)	Grape seed and grapefruit seed extract
de Moraes Crizel et al. (2018)	Olive pomace
Li et al. (2012)	Sugar beet pulp
Yang et al. (2016)	Distiller dried grains
Zhao et al. (2020)	Soy protein isolate and egg white
Garrido et al. (2014)	Soy protein from soy oil industry
da Silva e Silva et al. (2021)	Fish gelatine incorporated with palm oil and clove and oregano essential oils
de la Caba et al. (2019)	Sea food waste
Agro bio-based materials as substitutes for inorganic or petroleum-based compounds	
Nemes et al. (2020)	Seafood by-products (heads, gills, skin, trimmings); fruit and vegetable industries by-products (pumpkin seeds and peels, sunflower head, sisal waste, grapefruit, sour orange, pomegranate and eggplant peels)
Gaikwad et al. (2016)	Apple pomace
Costa-Trigo et al. (2019)	Chestnut wastes (leaves, prunings and burrs of chestnut trees and chestnut shells)
Banat (2019)	Olive pomace
Yang et al. (2018)	Wheat straw

(continued)

Table 2.1 (continued)

References	By-products
Garcia-Lomillo et al. (2014)	Wine pomace and grape seed
Biswas et al. (2020)	Essential oil from lime, nutmeg, eugenol, pimenta berry, rosemary, petitgrain, coffee, anise and trans-cinnamaldehyde
Avila et al. (2020)	Jaboticaba peel extracts
Bhardwaj et al. (2020)	Sugarcane bagasse, rice straw, wheat straw, corn stalks, barley residues and other lignocellulosic agro-waste
Porta et al. (2020)	Corn oil, chickpea flour, whey protein nanofibril, seed oil cakes
Agro bio-based for blending with fossil fuel-based polymers	
Szabo et al. (2020)	Tomato processing by-products extract
Sutivisedsak et al. (2012)	Cotton burr and cottonseed hull
Potrč et al. (2020), Zemljič et al. (2020)	Thyme, rosemary and cinnamon extracts were synthesized

2.2 Sustainability of Agro Bio-Based Compounds. A Brief Overview

The overview focuses on experimental papers and reviews as well.

In particular case studies in which the by-product is adopted as main ingredient (50%), and those in which it is secondary ingredient (37.5%) or additive (12.5%), are taken into account.

Among natural polymers, starch is a very promising material due to its wide abundance, biodegradability, low cost and annual renewability. Ren and coworkers studied the application of SPS film in food packaging (Ren et al. 2021).

In the same way, Lauer and colleagues analyzed the great potential of starch for packaging material both considering it as main ingredient and additive (Lauer and Smith 2020).

Hilliou and coworkers combined cheese whey and olive oil mill wastewater (OOMW) to obtain bioplastic films made of PHA. (Hilliou et al. 2016). One of the important advantages of these bio-packaging is their complete compostability at room temperature. Thus, in contrast to other commercially available biodegradable polymers and especially the mainly used PLA films petroleum-based biodegradable plastics, they do not require the use of industrial composting facilities (Bugnicourt et al. 2014). However, they present other disadvantages like the high difficulty to be converted into packages with conventional plastic processing techniques and their brittleness (Sanchez-Garcia et al. 2008).

Wu and colleagues experimented with bioactive edible films by incorporating different concentrations (5, 10, 15 and 20%) of tea polyphenol into pomelo peel flour in order to modify the functional film properties (Wu et al. 2019).

Jafarzadeh and coworkers experimented with plant extracts obtained from some tropical fruits (Avocado seed and peel, cocoa bean peel, coconut palm peel, mango kernel, potato peel) for biodegradable food packaging thanks to their natural origin and functionality (antimicrobial/antioxidant activity) (Jafarzadeh et al. 2020). Similar considerations apply to the study by Sá and colleagues in which a bacterial cellulose packaging film with antioxidant properties was realized (Sá et al. 2020) from cashew apple juice and addition of lignin (0–15% by weight) and cellulose nanocrystals (0–8% by weight).

Probably, the valorisation of wheat bran and straw by Souza Filho and coworkers could entail interesting sustainable implications considering that wheat is one of the most cultivated crops in the world (Souza Filho et al. 2020).

As far as biodegradable active food packaging, Adilah and colleagues used mango peel extract incorporated into fish gelatine films, demonstrating greater efficacy against free radicals (Adilah et al. 2018). Andres and coworkers studied the tomato pomace as a promising source of compounds with nutritional, antioxidant and antimicrobial potential (Andres et al. 2017).

da Rosa and colleagues studied olive leaf extract added to carrageenan film (da Rosa et al. 2019). In the paper by Nerantzis and coworkers, vine pruning was studied as a substitute for wood for the production of particleboard (Nerantzis and Tataridis 2006).

Sogut and colleagues evaluated the effect of grape seed extract (at 5, 10 and 15%) incorporated in the chitosan film considering its physico-mechanical properties, antioxidant and antimicrobial activities in order to show better shelf life for food packaged under vacuum refrigeration conditions (Sogut and Seydim 2018). Cerón Gómez developed two films based on pullulan and polylactic acid (PLA) derived from grape seed and grapefruit as antimicrobials (Cerón Gómez 2020). de Moraes Crizel and coworkers studied a type of packaging derived from flour and olive pomace flour microparticles in chitosan-based films (de Moraes Crizel et al. 2018). Li and colleagues explored a new strategy to make active packaging materials for food packaging from sugar beet pulps (Li et al. 2012).

Furthermore, Simkovic and coworkers studied holocelluloses from sugar beet and bagasse for film preparation (Simkovic et al. 2017). Yang and colleagues realized a new active packaging with interesting performance for pork meat (Yang et al. 2016).

Zhao and coworkers analyzed the effect of cinnamaldehyde concentration on the structural, physical and functional properties of edible composite films of soy protein isolate and egg white (EW) (Zhao et al. 2020). The soy protein as by-product of soy oil industry is also investigated by Garrido and colleagues in order to obtain a biodegradable packaging film (Garrido et al. 2014).

da Silva and coworkers applied the KNN algorithm to classify and select biodegradable packaging produced from fish gelatine incorporated with palm oil, clove and oregano essential oils (da Silva e Silva et al. 2021). Seafood wastes were also investigated by de la Caba and colleagues which highlighted the importance of reducing food waste (generated at the consumption) and food losses (generated along the chain) (de la Caba et al. 2019).

Some experimentations aim to integrate compounds extracted from by-products in usual technologies, such as food packaging by taking into account the recovering of seafood by-products for chitosan extraction and fruit and vegetable industries by-products for alginate and pectin (Nemes et al. 2020). Gaikwad and coworkers indicated that an active PVA/apple pomace film prepared by a melting process had great potential application in the consumption of apple pomace as a natural by-product of apple (Gaikwad et al. 2016).

Four kinds of waste from the industrial processing of chestnuts (*Castanea sativa*), namely leaves, pruned material and burrs from chestnut tree plus chestnut shells, were characterized to determine their content in polymers and thus their potential use in biorefinery (Costa-Trigo et al. (2019)). According to recently published data, the sustainable implications of the valorisation of this by-product could be relevant (FAOSTAT 2019), especially for the social impact on local producers in Mediterranean Areas (Spain, Turkey, Italy, Portugal and Greece are respectively the main producers).

In the same way, Banat investigated the use of renewable natural fibres such as olive stone flour as new organic fillers in polymer matrices (Banat 2019).

Wheat straw has gained great interest as a filler in green composites, but some studies on PLA/wheat straw composites have reported limited reductions in physical performance. The study by Yang and coworkers reports on a novel method for the production of cellulose nanofibrils by solid-state machining to produce PLA/wheat straw composites (Yang et al. 2018).

Garcia-Lomillo and colleagues showed the great antioxidant and bacteriostatic activity of substances extracted from whole wine pomace and grape seeds, which could have interesting applications in active food packaging films (Garcia-Lomillo et al. 2014).

Biswas and coworkers showed that some essential oils in cellulose ester films for food packaging have at least four valuable characteristics: higher flexibility, lower water vapour permeability, variable opacity and antimicrobial activity (Biswas et al. 2020).

Obviously, the sustainability of these bio-based compounds is linked to their production process that starting from the primary stage of crop cultivation, should be compliant with the best available technologies in terms of water and energy saving.

In order to individuate natural fibres for substituting inorganic compounds, Missio and colleagues experimented the combination of cellulose nanofibrils (CNF) and condensed tannins from *Acacia mearnsii* for the development of hybrid, functional films (Missio et al. 2020). Nevertheless, no information about sustainable aspects linked to this possible packaging solution were provided by the authors. Some problems, for example, could occur in considering the negative impacts that *A. mearnsii* growing could have in terms of biodiversity reduction and high-water requirement according to its areas of origin and diffusion.

The same considerations could be done for the experimentation of Avila and coworkers that showed the functional activity of jaboticaba peel extracts due to high total phenolic content, antioxidant activity and total anthocyanin (Avila et al. 2020). Bhardwaj and colleagues analyzed the potential of sustainable packaging solutions coming from agricultural by-products/wastes such as sugarcane bagasse, rice straw, wheat straw, corn stalks, barley residues and other lignocellulosic compounds (Bhardwaj et al. 2020).

Porta and coworkers reviewed studies about bio-polymers potentially usable in the areas of both food and non-food packaging (Porta et al. 2020). As concerning bio-based material used in blending with fossil fuel-based polymers, Banat involved the use of olive pomace to obtain a biodegradable composite material filled with PHBV of biodegradable olive pomace (Banat 2019).

Nemes and colleagues experimented with active films prepared from PVA mixed with itaconic acid and with chitosan, enriched with an extract of by-products of tomato processing in order to develop new bioactive formations for the food packaging. In evaluating this solution, it is important to point out that these and similar agro-food by-products (such as olive pomace, grape marc, etc.) are produced for a short period of the year (Nemes et al. 2020).

Sutivisedsak and coworkers considered two by-products deriving from cotton, namely cotton burr and cottonseed hull, which were used as filling materials in composites of PLA and LDPE and PHA (Sutivisedsak et al. 2012). In this case, sustainability should be well assessed in accordance with the high impacts of cotton production (especially for the great crop water requirement and the use of fertilizers and pesticides).

Potrč and colleagues and Zemljič and coworkers reported the great potential of PE and PP packaging foils, coated by chitosan and with embedded thyme, rosemary and cinnamon extracts, as active (antioxidant and antimicrobial) packaging in the food industry (Potrč et al. 2020; Zemljič et al. 2020).

2.3   Assessing Sustainability to Bio-Based Materials Coming from Agro-Food By-Products

The Life Cycle Assessment is a worldwide well-known methodology and recognized as valid instrument to assess sustainability. It was standardized by the International Standard Organization (ISO 14040:2006; ISO 14044:2006) and adopted in the main environmental certification schemes as for products, services and organizations as well (Benini et al. 2014;EC 2013, 2018; Ojala et al. 2016).

The LCA considers the product as a system of products in which all material and energy flows are taken into account as input, while the emissions in air, soil and water, as well as, wastes and by-product are accounted as output of the system (Curran 2012; Ekvall and Weidema 2004; Ekvall et al. 2016; JRC-IES 2010).

In this ambit, an agro-food by-product is normally generated in a system in which the aim is to produce the main agro-food product destined directly to consumers as foodstuff (e.g. fruits and vegetables) or indirectly as raw material for another agro-food process (olives for olive oil production or milk for dairies etc.). For these reasons, in assessing sustainability it needs to calculate the fraction of impacts that could be linked to that by-product with respect to the part addressed to the main agro-food product. In the LCA methodology, this issue is known as "allocation procedure" and it is recommended to be set according to criteria able to represent the importance, such as in terms of weight or economic value, of the different co-products obtained from the same system.

In agro-food systems, the percentage in weight of by-products with respect to the main product could be high and that could mislead the analysis if an allocation procedure based on weight is carried out. This is true, for example in the case of olive oil processing, in which olive pomace represents almost 50% of olives (and the main product, olive oil about 20%). In the same way, in the system of production of fresh meat, the by-products and waste of slaughtering cover over 50% of the cattle live weight or whey which could be until the 85% of the milk respect to the cheese. Then an allocation procedure based on economic value could appear more accurate.

It is also true, that, at the moment, as shown in the above-mentioned overview, except for some commercial processes such as PLA production, almost all attempts of exploitation of agro-food by-product for obtaining bio-based compounds are on an experimentation level. So, due to the fact that there isn't a real market for these by-products, it is not easy to individuate an allocation factor able to represent the real value of the by-product with respect to the main agro-food product.

Furthermore, it is also worth noting that, in a life cycle perspective, as concerning some impact categories such as the Global Warming Potential (GWP), in some cases the destination of the by-product as raw material for biorefinery worse the overall picture of the impacts. For example, as for straw or pruning residues, if they are let on field and grinded or buried into soil, there will be, without considering erosion and leaching, various positive effects will derive from this, such as an immediate increase in the stock of SOC (soil organic carbon) (Bird et al. 2011) which subsequently, with the decomposition, can be partially humified becoming a more stabilized share of SOC (more durable form of carbon storage) and partly mineralized providing more nourishment to the soil microbiota, which determines an increase in soil fertility (Guggenberger 2005) and releasing nutrients for plants that will require less fertilizers intake (Bird et al. 2011; Guggenberger 2005), and advantages in terms of GHG could be accounted as carbon dioxide (CO_2)-equivalent for the relative impact categories such as GWP.

Otherwise, if they are brought away from the system and considered in another process the effects of carbon uptake and sequestration will be cancelled.

These considerations entail that the problem of sustainability of bio-based packaging materials, in a life cycle perspective, should be faced by a system expansion with a consequential approach (Ekvall et al. 2016). In detail, a way could be comparing the avoided impacts linked to the substitution of petroleum-based packaging material with the bio-based ones.

2.3.1   Comparison of GHG Emissions of Petroleum-Based Polymers with Biodegradable Packaging Materials

According to what asserted above, the following simulation aims to formulate a reliable hypothesis, by using the LCA methodology, about how to address the environmental performances of the agro-food bio-based compounds. The analysis takes into consideration the Global Warming impact categories but, obviously, the approach could be applied to all the impact categories both environmental and economic or social as well.

The comparison involves nine by-products: olive pit (OP), straw (S), wine pomace (WP), whey (W), cellulose fibres from forestry (CF), coconut husk (CH), waste from marine fishery (MF), apple pomace (AP), and sugar beet pulp (SBP), compared with the most used petroleum-based polymers: polyethylene (PE) and polypropylene (PP) and polylactide (PLA) obtained from corn.

The life cycle modelling of the different scenarios was carried out using the Gabi software[1] and the databases Sphera, Plastics Europe and Ecoinvent (Frischknecht and Jungbluth 2007; IKP and PE 2002; Nemecek and Kägi 2007; Pfister et al. 2015; Weidema et al. 2013). As for the life cycle of straw, olive pit and wine pomace data derive from direct data collection and previous elaboration (Cappelletti et al. 2014; Russo et al. 2016, 2019). While the Ecoinvent processes are adopted for the other by-products considering the allocation factors shown in Table 2.2. In particular, as for whey, the Ecoinvent process of cow milk cheese production was considered by applying an allocation factor of 0.2. In the same way, as far as waste from marine fishery, the Ecoinvent wild-fish process was taken into account and an allocation factor of 0.5 was accounted. The Ecoinvent database was also used for processes of PLA production from corn and PP e PE-LD granulates.

Figure 2.1 shows the results, in terms of kg of CO_2-equivalent to the GWP of 1 kg of each bio-based material compared with 1 kg of PP, LDPE and PLA granulate. The GWP is subdivided between GWP fossil as positive value of CO_2-equiv. (carbon emission) and GWP biogenic accounted as negative value (carbon up-take).

Due to the lack of information about processes for obtaining packaging material, it is not possible to compare the environmental performances of the final product, but, despite that, it is possible to formulate some considerations and hypotheses about how the comparison should be carried out.

Figure 2.1 confirms the advantages of agro-food by-products in terms of GWP biogenic due to the carbon up-take. In the same time, PLA obtained from corn presents a high value of GWP fossil that is however balanced by GWP biogenic but probably could be reduced by using other starchy agro-food by-products rather than corn, which, in spite of its low price, could be used for food and feed.

---

[1]This software is available at http://www.gabi-software.com.

Table 2.2 Data sources and allocation factors (AF) of agro-food by-products

By-product	Main agro-food product	AF	Source of data
OPF	Olive oil	1	Our elaboration
S	Wheat	1	Our elaboration
WP	Wine	1	Our elaboration
W	Cheese	0.2	Cheese production, ecoinvent 3.5
CF	Wood from forestry	1	Cellulose fibres, ecoinvent 3.5
CH	Coconut	1	Coconut husk, ecoinvent 3.5
MF	Marine fish	0.5	Marine fish, ecoinvent 3.5
AP	Apple juice	1	Apple production, ecoinvent 3.5
SBP	Sugar	1	Beet sugar production, ecoinvent 3.5
PE-LD		1	Polyethylene low density granulate ecoinvent 3.5
PP		1	Polypropylene granulate ecoinvent 3.5
PLA		1	Polylactide ecoinvent 3.5

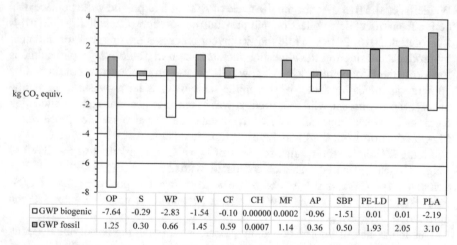

	OP	S	WP	W	CF	CH	MF	AP	SBP	PE-LD	PP	PLA
☐GWP biogenic	-7.64	-0.29	-2.83	-1.54	-0.10	0.00000	0.0002	-0.96	-1.51	0.01	0.01	-2.19
▣GWP fossil	1.25	0.30	0.66	1.45	0.59	0.0007	1.14	0.36	0.50	1.93	2.05	3.10

Fig. 2.1 GWP fossil and biogenic of 1 kg of agro-food by-product compared to 1 kg of PP, LDPE and PLA

However, in order to assess the sustainability of using agro-food by-products as bio-based material for food packaging in substitution of petroleum-based compounds such as PP or LDPE the relative impacts of the process for obtaining packaging material from the formers (e.g. biodegradable film) should be reduced according to the avoided impact deriving from the non-production of petroleum-based packaging material.

According to the Sphera and Plastics Europe Databases, the GWP fossil of PE and PP films is respectively 2.37 and 2.11 kg of CO_2 equivalents, these values

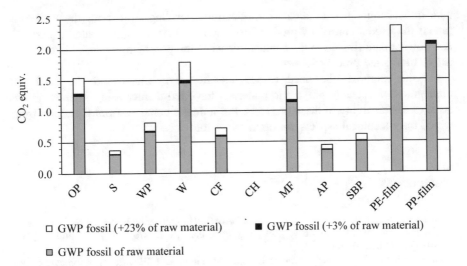

Fig. 2.2 GWP fossil of 1 kg of agro-food by-product worsened for a range from 3 to 23% compared to 1 kg of PP and LDPE film

represent 23% and 3% plus of the impact of raw material, respectively. If the same range of 3–23% of worsening is applied to the other bio-based raw materials, the situation is that described in Fig. 2.2. In all situations, the impacts of the agro bio-based compounds are lower than the ones of petroleum-based materials. Anyhow, according to the hypothesis of considering the avoided impacts of plastics production, the sustainability of the agro bio-based could be acceptable if the value of the impact category doesn't overcome the double amount referred to the petroleum-based compound.

2.4 Sustainability of Agro Bio-Based Components for Packaging Materials: An Open Issue

As concerning sustainability of agro-food product, it is worth noting that the agricultural phase, is in many studies, indicates as the main responsible of the main environmental and social impacts (Cucurachi et al. 2019; Salomone et al. 2015). This entails the importance of reducing the impacts of the overall system by valorising the potential of by-products as bio-based material for food packaging. However, it is also true that the life cycle of those packaging materials should be well investigated in order to evaluate the real sustainability of by-product valorisation.

On the other hand, when food waste occurs it means that consumption of material and energy resources and emissions linked to the main product were generated uselessly. In this ambit, the research in agro bio-based compounds should take into account the sustainable implications deriving from both the upstream (about the production of by-product from which the packaging material derives) and downstream (linked to the characteristics of the packaging material in terms of active functions) processes.

At the moment, the need to respect the environment and the need for biodegradable versatile polymeric materials have led to increasing interest on the issue of sustainability of these products, but it doesn't emerge from literature, from which only technical aspects are often taken into account.

References

Adilah AN, Jamilah B, Noranizan MA, Hanani ZAN (2018) Utilization of mango peel extracts on the biodegradable films for active packaging. Food Pack Shelf Life 16:1–7. https://doi.org/10.1016/j.fpsl.2018.01.006

Andres AI, Petron MJ, Delgado-Adamez J, Lopez M, Timon M (2017) Effect of tomato pomace extracts on the shelf-life of modified atmosphere-packaged lamb meat. J Food Proc Preserv 41 (4):e13018. https://doi.org/10.1111/jfpp.13018

Avila LB, Vaz Fontes MR, da Rosa Zavareze E, Costa Moraes C, Machado Morais M, Silveira da Rosa G (2020) Recovery of bioactive compounds from jaboticaba peels and application into zein ultrafine fibers produced by electrospinning. Polym 12(12):2916. https://doi.org/10.3390/polym12122916

Banat R (2019) Olive pomace flour as potential organic filler in composite materials: a brief review. Am J Polym Sci 9(1):10–15. https://doi.org/10.5923/j.ajps.20190901.02

Benini L, Mancini L, Sala S, Manfredi S, Schau EM, Pant R (2014) Normalization method and data for environmental footprints. Report EUR 26842 EN. Ispra, Italy

Bhardwaj A, Alam T, Sharma V, Alam MS, Hamid H, Deshwal GK (2020) Lignocellulosic agricultural biomass as a biodegradable and eco-friendly alternative for polymer-based food packaging. J Pack Technol Res 4(2):205–216. https://doi.org/10.1007/s41783-020-00089-7

Bird N, Cowie A, Cherubini F, Jungmeier G (2011) Using a life cycle assessment approach to estimate the net greenhouse gas emissions of bioenergy. IEA Bioenergy, Strategic Report, ExCo:2011:03

Biswas A, do Socorro Rocha Bastos M, Furtado RF, Kuzniar G, Boddu V, Cheng HN (2020) Evaluation of the properties of cellulose ester films that incorporate essential oils. Int J Polym Sci 2020. https://doi.org/10.1155/2020/4620868. Article ID 4620868

Brennan L, Langley S, Verghese K, Lockrey S, Ryder M, Francis C, Phan-Le NT, Hill A (2020) The role of packaging in fighting food waste: a systematised review of consumer perceptions of packaging. J Clean Prod. https://doi.org/10.1016/j.jclepro.2020.125276. 125276

Bugnicourt E, Cinelli P, Lazzeri A, Alvarez V (2014) Polyhydroxyalkanoate (PHA): review of synthesis, characteristics, processing and potential applications in packaging. Expr Polym Lett 8(11):791–808. https://doi.org/10.3144/expresspolymlett.2014.82

Cappelletti GM, Ioppolo G, Nicoletti GM, Russo C (2014) Energy requirement of extra virgin olive oil production. Sustainability 6(8):4966–4974. https://doi.org/10.3390/su6084966

Cerón Gómez MM (2020) Development of bio-based active packaging material evaluating grapefruit seed and grape seed extract as antimicrobials. Dissertation, Escuela Agrícola Panamericana, Zamorano

Chisenga SM, Tolesa GN, Workneh TS, Owusu-Kwarteng J (2020) Biodegradable food packaging materials and prospects of the fourth industrial revolution for tomato fruit and product handling. Int J Food Sci 2020:1–17. https://doi.org/10.1155/2020/8879101

Conte A, Cappelletti GM, Nicoletti GM, Russo C, Del Nobile MA (2015) Environmental implications of food loss probability in packaging design. Food Res Int 78:11–17. https://doi.org/10.1016/j.foodres.2015.11.015

Costa-Trigo I, Otero-Penedo P, Outeirino D, Paz A, Dominguez JM (2019) Valorization of chestnut (Castanea sativa) residues: characterization of different materials and optimization of the acid-hydrolysis of chestnut burrs for the elaboration of culture broths. Waste Manag 87:472–484. https://doi.org/10.1016/j.wasman.2019.02.028

Cucurachi S, Scherer L, Guinée J, Tukker A (2019) Life cycle assessment of food systems. One Earth 1(3):292–297. https://doi.org/10.1016/j.oneear.2019.10.014

Curran MA (2012). Life cycle assessment handbook: a guide for environmentally sustainable products. Wiley, Hoboken

da Rosa GS, Vanga SK, Gariepy Y, Raghavan V (2019) Development of biodegradable films with improved antioxidant properties based on the addition of carrageenan containing olive leaf extract for food packaging applications. J Polym Environ 28(1):123–130. https://doi.org/10.1007/s10924-019-01589-7

da Silva e Silva N, de Souza Farias F, dos Santos Freitas MM, Pino Hernández EJG, Dantas VV, Enê Chaves Oliveira M, Joele MRSP, de Fátima Henriques Lourenço L (2021) Artificial intelligence application for classification and selection of fish gelatin packaging film produced with incorporation of palm oil and plant essential oils. Food Pack Shelf Life 27. https://doi.org/10.1016/j.fpsl.2020.100611. 100611

de la Caba K, Guerrero P, Trung TS, Cruz-Romero M, Kerry JP, Fluhr J, Maurer M, Kruijssen F, Albalat A, Bunting S, Burt S, Little D, Newton R (2019) From seafood waste to active seafood packaging: an emerging opportunity of the circular economy. J Clean Prod 208:86–98. https://doi.org/10.1016/j.jclepro.2018.09.164

de Moraes Crizel T, de Oliveira Rios A, Alves VD, Bandarra N, Moldão-Martins M, Hickmann Flôres S (2018) Active food packaging prepared with chitosan and olive pomace. Food Hydrocoll 74:139–150

EC (2013) Recommendations. Commission Recommendation of 9 April 2013 on the use of common methods to measure and communicate the life cycle environmental performance of products and organizations (Text with EEA relevance) (2013/179/EU). Off J Eur Union L124:1–210

EC (2018) European Commission, Environmental Footprint Guidance document—guidance for the development of Product Environmental Footprint Category Rules (PEFCRs), version 6.3

Ekvall T, Azapagic A, Finnveden G, Rydberg T, Weidema BP, Zamagni A (2016) Attributional and consequential LCA in the ILCD handbook. Int J Life Cycle Assess 21(3):293–296. https://doi.org/10.1007/s11367-015-1026-0

Ekvall T, Weidema BP (2004) System boundaries and input daa in consequential life cycle inventory analysis. Int J Life Cycle Assess 9(3):161–171. https://doi.org/10.1007/BF02994190

FAOSTAT (2019) Crops statistics. The Food and Agriculture Organization of the United Nations, Rome. Available http://www.fao.org/faostat/en/#data. Accessed 17 April 2021

Frischknecht R, Jungbluth N (eds) (2007) Overview and methodology—ecoinvent report No. 1. Swiss Centre for Life Cycle Inventories, Dubendorf. Available http://www.pre-sustainability.com/download/manuals/EcoinventOverviewAndMethodology.pdf. Accessed 17 April 2021

Gaikwad KK, Lee JY, Lee YS (2016) Development of polyvinyl alcohol and apple pomace bio-composite film with antioxidant properties for active food packaging application. J Food Sci Technol 53(3):1608–1619. https://doi.org/10.1007/s13197-015-2104-9

Garcia-Lomillo J, Gonzalez-SanJose ML, Del Pino-Garcia R, Rivero-Perez MD, Muniz-Rodriguez P (2014) Antioxidant and antimicrobial properties of wine byproducts and their potential uses in the food industry. J Agric Food Chem 62, 52:12595–12602. https://doi.org/10.1021/jf5042678

Garrido T, Etxabide A, LecetaI CS, de la Caba K, Guerrero P (2014) Valorization of soya by-products for sustainable packaging. J Clean Prod 64:228–233. https://doi.org/10.1016/j.jclepro.2013.07.027

Guggenberger G (2005). Humification and mineralization in soils. In: Varma A, Buscot F (eds) Microorganisms in soils: roles in genesis and functions. Soil biology, vol 3. Springer, Berlin, Heidelberg

Hilliou L, Machado DC, Oliveira SS, Gouveia AR, Reis MAM, Campanari S, Villano M, Majone M (2016) Impact of fermentation residues on the thermal, structural, and rheological properties of polyhydroxy (butyrate-co-valerate) produced from cheese whey and olive oil mill wastewater. J Appl Polym Sci 133(2):42818. https://doi.org/10.1002/app.42818

IKP, PE (2002) GaBi 4—software-system and databases for life cycle engineering. Echterdingen, Stuttgart

ISO 14040 (2006) Environmental management—life cycle assessment—principles and framework ISO 14040:2006. International Organization for Standardization, Geneva

ISO 14044 (2006) Environmental management—life cycle assessment—requirements and guidelines. ISO 14044:2006. International Organizationfor Standardization, Geneva

Jafarzadeh S, Jafari SM, Salehabadi A, Nafchi AM, Uthaya Kumar US, Khalil HPSA (2020) Biodegradable green packaging with antimicrobial functions based on the bioactive compounds from tropical plants and their by-products. Trends Food Sci Technol 100:262–277. https://doi.org/10.1016/j.tifs.2020.04.017

JRC-IES (2010) International reference life cycle data system (ILCD) handbook—general guide for life cycle assessment—detailed guidance. Publications Office of the European Union, Luxembourg

Lauer MK, Smith RC (2020) Recent advances in starch-based films toward food packaging applications: physicochemical, mechanical, and functional properties. Compr Rev Food Sci Food Saf 19(6):3031–3083. https://doi.org/10.1111/1541-4337.12627

Li W, Coffin DR, Jin TZ, Latona N, Liu CK, Liu B, Zhang J, Liu L (2012) Biodegradable composites from polyester and sugar beet pulp with antimicrobial coating for food packaging. J Appl Polym Sci 126(S1):E362–E373. https://doi.org/10.1002/app.36885

Missio AL, Mattos BD, Otoni CG, Gentil M, Coldebella R, Khakalo A, Gatto DA, Rojas OJ (2020) Cogrinding wood fibers and tannins: surfactant effects on the interactions and properties of functional films for sustainable packaging materials. Biomacromolecul 21(5):1865–1874. https://doi.org/10.1021/acs.biomac.9b01733

Nemecek T, Kägi T (2007) Life cycle inventories of agricultural production systems. Data v2.0. Ecoinvent report No. 15.

Nemes SA, Szabo K, Vodnar DC (2020) Applicability of agro-industrial by-products in intelligent food packaging. Coat 10(6):550. https://doi.org/10.3390/coatings10060550

Nerantzis ET, Tataridis P (2006) Integrated enology-utilization of winery by-products into high added value products. e-J Sci Technol 1(3):79–89

Ojala E, Uusitalo V, Virkki-Hatakka T, Niskanen A, Soukka R (2016) Assessing product environmental performance with PEF methodology: reliability, comparability, and cost concerns. Int J Life Cycle Assess 21(8):1092–1105. https://doi.org/10.1007/s11367-016-1090-0

Payen S, Basset-Mens C, Colin F, Roignant P (2017) Inventory of field water flows for agri-food LCA: critical review and recommendations of modelling options. Int J Life Cycle Assess 23:1331–1350. https://doi.org/10.1007/s11367-017-1353-4

Pfister S, Vionnet S, Levova T, Humbert S (2015) Ecoinvent 3: assessing water use in LCA and facilitating water footprinting. Int J Life Cycle Assess 21(9):1349–1360. https://doi.org/10.1007/s11367-015-0937-0

Porta R, Sabbah M, Di Pierro P (2020) Biopolymers as food packaging materials. Int J Mol Sci 21(14):4942. https://doi.org/10.3390/ijms21144942

Potrč S, Sterniša MAO, Smole Možina SAO, Knez Hrnčič M, Fras Zemljič L (2020) Bioactive characterization of packaging foils coated by chitosan and polyphenol colloidal formulations. Int J Mol Sci 21(7):2610. https://doi.org/10.3390/ijms21072610

Ren Y, Wu Z, Shen M, Rong L, Liu W, Xiao W, Xie J (2021) Improve properties of sweet potato starch film using dual effects: combination Mesona chinensis Benth polysaccharide and sodium carbonate. LWT Food Sci Technol 140. https://doi.org/10.1016/j.lwt.2020.110679. 110679

Russo C, Cappelletti GM, Nicoletti GM, di Noia AE, Michalopoulos G (2016) Comparison of European olive production systems. Sustainability 8(8):825. https://doi.org/10.3390/su8080825

Russo C, Cappelletti GM, Nitkiewicz T (2019) Best environmental practices in agricultural sector: comparative LCA study of Italian and polish wheat cultivation. In: Salerno-Kochan R (ed) Quality science in the face of the challenges of innovative economy and sustainable development Nauki o zarządzaniu i jakości wobec wyzwań zrównoważonego rozwoju = Management and Quality Studies Facing Challenges of Sustainable Development. Instytut Technologii Eksploatacji—Państwowy Instytut Badawczy, Radom, Kraków, Poland

Sá N, Mattos A, Silva LM, Brito E, Rosa M, Azeredo H (2020) From cashew byproducts to biodegradable active materials: bacterial cellulose-lignin-cellulose nanocrystal nanocomposite films. Int J Biol Macromolecul 161:1337–1345. https://doi.org/10.1016/j.ijbiomac.2020.07.269

Salomone R, Cappelletti GM, Malandrino O, Mistretta M, Neri E, Nicoletti GM, Notarnicola B, Pattara C, Russo C, Saija G (2015) Life cycle assessment in the olive oil sector. In: Notarnicola B, Salomone R, Petti L, Renzulli P, Roma R, Cerutti A (eds) Life cycle assessment in the agri-food sector. Springer, Cham. https://doi.org/10.1007/978-3-319-11940-3_2

Sanchez-Garcia MD, Gimenez E, Lagaron JM (2008) Morphology and barrier properties of solvent cast composites of thermoplastic biopolymers and purified cellulose fibers. Carbohydr Polym 71(2):235–244. https://doi.org/10.1016/j.carbpol.2007.05.041

Simkovic I, Kelnar I, Mendichi R, Bertok T, Filip J (2017) Composite films prepared from agricultural by-products. Carbohydr Polym 156:77–85. https://doi.org/10.1016/j.carbpol.2016.09.014

Sogut E, Seydim AC (2018) The effects of Chitosan and grape seed extract-based edible films on the quality of vacuum packaged chicken breast fillets. Food Pack Shelf Life 18:13–20. https://doi.org/10.1016/j.fpsl.2018.07.006

Souza Filho PF, Zamani A, Ferreira JA (2020) Valorization of wheat byproducts for the co-production of packaging material and enzymes. Energies 13(6):1300. https://doi.org/10.3390/en13061300

Sutivisedsak N, Cheng HN, Dowd MK, Selling GW, Biswas A (2012) Evaluation of cotton byproducts as fillers for poly(lactic acid) and low density polyethylene. Ind Crops Prod 36(1):127–134. https://doi.org/10.1016/j.indcrop.2011.08.016

Szabo K, Teleky BE, Mitrea L, Călinoiu LF, Martău GA, Simon E, Varvara RA, Vodnar DC (2020) Active packaging—poly(Vinyl alcohol) films enriched with tomato by-products extract. Coatings 10(2):141. https://doi.org/10.3390/coatings10020141

Valdes A, Mellinas AC, Ramos M, Garrigos MC, Jimenez A (2014) Natural additives and agricultural wastes in biopolymer formulations for food packaging. Front Chem 2:6. https://doi.org/10.3389/fchem.2014.00006

Varžinskas V, Markevičiūtė Z (2020) Sustainable food packaging: materials and waste management solutions. Environ Res Eng Manag 76(3):154–164. https://doi.org/10.5755/j01.erem.76.3.27511

Weidema BP, Bauer C, Hischier R, Mutel C, Nemecek T, Reinhard J, Vadenbo CO, Wernet G (2013) Overview and methodology. Data quality guideline for the ecoinvent database version 3. Ecoinvent Report 1(v3). The ecoinvent Centre, St. Gallen. Available https://www.ecoinvent.org/files/dataqualityguideline_ecoinvent_3_20130506.pdf. Accessed 17 April 2021

Wu H, Lei Y, Zhu R, Zhao M, Lu J, Xiao D, Jiao C, Zhang Z, Shen G, Li S (2019) Preparation and characterization of bioactive edible packaging films based on pomelo peel flours incorporating tea polyphenol. Food Hydrocoll 90:41–49. https://doi.org/10.1016/j.foodhyd.2018.12.016

Yang HJ, Lee JH, Won M, Song KB (2016) Antioxidant activities of distiller dried grains with solubles as protein films containing tea extracts and their application in the packaging of pork meat. Food Chem 196:174–179. https://doi.org/10.1016/j.foodchem.2015.09.020

Yang S, Bai S, Wang Q (2018) Sustainable packaging biocomposites from polylactic acid and wheat straw: enhanced physical performance by solid state shear milling process. Compos Sci Technol 158:34–42. https://doi.org/10.1016/j.compscitech.2017.12.026

Zemljič LF, Plohl O, Vesel A, Luxbacher T, Potrč S (2020) Physicochemical characterization of packaging foils coated by Chitosan and polyphenols colloidal formulations. Int J Mol Sci 21 (2):495. https://doi.org/10.3390/ijms21020495

Zhang H, Hortal M, Dobon A, Bermudez JM, Lara-Lledo M (2015) The effect of active packaging on minimizing food losses: life cycle assessment (LCA) of essential oil component-enabled packaging for fresh beef. Pack Technol Sci 28(9):761–774. https://doi.org/10.1002/pts.2135

Zhao X, Mu Y, Dong H, Zhang H, Zhang H, Chi Y, Song G, Li H, Wang L (2020) Effect of cinnamaldehyde incorporation on the structural and physical properties, functional activity of soy protein isolate-egg white composite edible films. J Food Proc Preserv 45(2):e15143. https://doi.org/10.1111/jfpp.15143

Chapter 3
New Eco-Friendly Packaging Strategies Based on the Use of Agri-Food By-Products and Waste

Abstract The wide use of petroleum-derived plastics and their negative impact on the environment require deep research on biodegradable materials obtained from renewable resources. The preserved food industry must sustain increasing costs for treating solid and liquid wastes. In fact, the use of these materials for animal feed or fertilizer without pre-treatments is not easy because of the intolerance of some animals to some waste components and the known germination inhibition properties of many polyphenols. The use of this material to develop innovative biodegradable packaging could represent an exciting opportunity. This chapter gives an overview of the leading research related to agri-food by-products and industrial food wastes to realize biodegradable food packaging.

Keywords Biopolymers · Food packaging · Bioplastics · Compostable material · Microbial processing · Chemical extraction · Extrusion-cooking processing

Abbreviations

T_{gel}	Gelatinization temperature
LDPE	Low density polyethylene
T_m	Melting temperature
PHAs	Polyhydroxyalkanoate
PLA	Poly-lactic acid
PVC	Polyvinyl chloride
SC-CO$_2$	Supercritical carbon dioxide
SFE	Supercritical fluid extraction
STI	Sustainable Technologies Initiative
TPS	Thermoplastic starch

© The Author(s), under exclusive license to Springer Nature Switzerland AG 2021
T. De Pilli et al., *Sustainable Innovations in Food Packaging*, Chemistry of Foods,
https://doi.org/10.1007/978-3-030-80936-2_3

3.1   Background and Problem Statement of Eco-Friendly Food Packaging

The new policies based on industrial development and economic growth require converting all production activities to guarantee environmental sustainability. This ambitious goal involves the need to reduce and replace plastic materials, which derive from petroleum due to the negative impact on the environment and the non-renewability of the source from which they are produced (Weber et al. 2002). In 2018 the global plastic production was almost 360 million tons (European Bioplastics 2020), while bioplastics production was 0.56% of the world's plastic production (Prieto 2016). In recent years, this request has given impetus to the study and research of biodegradable materials deriving from renewable sources (Laufenberg et al. 2003; Denavi et al. 2009). One of the leading industries that use plastic material to make their products is that of food packaging. Therefore, the conversion to an eco-sustainable industry requires the use of organic polymers (as protein, polysaccharides, etc.) from naturally renewable sources such as by-products or waste from agricultural production or the food industry (Salgado et al. 2008), that could get a circular economy (Geueke et al. 2018; Imre et al. 2019).

The conversion of production of food packaging through the substitution of plastic material with biodegradable materials from renewable sources arises both from the difficulty of disposing of large volumes of high environmental impact material and from the problems in the recovery and recycling of packaging after their use because of food contamination and high cost that requires this activity.

Packaging made using biopolymers extracted from agricultural and food industry by-products and waste do not increase the amount of carbon dioxide in the atmosphere, require low production costs and are economically sustainable. Furthermore, their production process involves the economic enhancement of materials that should be disposed of. In addition, this new use also involves a drastic reduction in the costs of disposal of waste materials by simplifying their management activity.

Large amounts of food waste come from food industries like canning industries (such as those that process artichokes, asparagus, tomatoes, etc.) and beverage industries such as winemaking.

Functional compounds can be extracted from these wastes, such as polyphenols and flavonoids, using eco-friendly techniques such as extraction with supercritical fluids (SFE) (Di Mauro et al. 2002). Furthermore, the material obtained after extraction with this method could produce new biodegradable food packaging.

The wine industry is one of the most widespread beverage industries with a higher production volume than other companies belonging to the same category. The world production of grapes (*Vitis vinifera*) for transformation into wine has reached a quota of 76 million tons. Numerous by-products and waste materials are generated during the production of wine: stalks, pomace (skins and seeds), lees, stems, perlite and diatomaceous earth residues, cellulosic filters and bentonite residues. The marc and lees are by-products destined to produce alcoholic beverages produced by distillation, which produces exhausted marc as waste. Most oenological waste, except grape stems, which are sometimes used as fertilizers without any pre-treatment, must be treated and disposed of as special waste, increasing the disposal costs and, therefore, wine production costs.

Many studies are currently performed to find and optimize the technologies, decrease production costs and improve innovative biodegradable food packaging (Ötles 2004).

The primary function of a material used for the production of food packaging is to contain and protect the food from the external environment while preserving the quality of the packaged product. This function mainly depends on the mechanical and diffusional properties of the material used for packaging. Therefore, studies on biodegradable packaging have focused on all those factors and technologies that can improve these properties and on food-packaging interactions during the shelf life that can change the performance of packaging materials over time (Siracusa et al. 2008).

There are many studies and investigational applications of biodegradable packaging applied to food packaging alone or with the addition of antimicrobials, chelators, antioxidants. The results of these studies have shown that these packaging materials are not only able to increase the quality and safety of packaged foods, slowing deterioration and prolong the shelf life, but also to confer characteristics of colour and taste better (Cutter 2006; Denavi et al. 2009; Suyatma et al. 2004).

All over the world are very numerous research centres (NatureWorks LLC, Novamont, Plantic Technologies Ltd., Rodenburg, Innovia Films and Procter and Gamble) who have focused their studies on the development of new biodegradable materials produced from renewable sources such as agricultural and food industry waste and by-products.

An example of this researches is the STI project or Sustainable Technologies Initiative, developed by the collaboration between the Imperial College of London and Pira International, McCain Foods, Greenvale AP, Organic Farm Alimenti and HL Hutchinson. The results of this project highlight the production of trays for packaging fruit and vegetables using starch extracted from the scraps of potato processing.

Despite the wide availability and simplicity of the extraction process from by-products and processing waste from the food industry, starch cannot be used to produce food packaging without undergoing profound modifications to improve its mechanical properties and barrier to water vapour. These modifications can be carried out with physical or chemical means or by adding polymeric additives or plasticizers. The addition to starch of biopolymers like fibre and protein may improve its mechanical and barrier properties.

3.2   The Biopolymers Are the Key to Realize Biodegradable Food Packaging

The main factors for producing and developing biodegradable materials are the polymers that make up the biological materials, called used biopolymers (Van de Velde and Kiekens 2002). These compounds are characterized by their ability to be

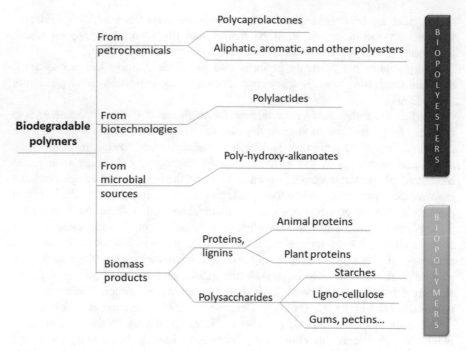

Fig. 3.1 Classification of the main biodegradable polymers adapted from Avérous and Pollet (2012)

metabolized easily and quickly by microorganisms, which can be defined as biodegradable compounds. These biological components were classified by manufacturing different criteria: type of chemical structure, origin, technological production and extraction (Fig. 3.1), application and cost–benefit ratio (Davidović and Savić 2010).

3.2.1 Biopolymers Extracted from Agriculture Crops

The main agro-polymers presented in this chapter are polysaccharides and proteins. Both biopolymers realize the multiphase packaging. However, polysaccharides can be considered the most abundant macromolecules in the biosphere.

3.2.1.1 Starch

Starch represents the most important energy reserve of vegetables and is present in cereals (wheat, rice, corn, barley, oats, etc.), tubers (potatoes, cassava, etc.), legumes, etc. Starch consists of two polymers constituted by glucose. Amylose is a linear polymer in which the glucose units are linked by α $(1 \rightarrow 4)$ linkage. Amylopectin is a branched polymer and glucose units are linearly linked to each other by the α $(1 \rightarrow 4)$ linkage, while the branches are characterized by the α $(1 \rightarrow 6)$ linkage. Amylose and amylopectin are arranged in such a way as to form a structure similar to granules, whose shape and size depends not only on the botanical origin of the vegetable plant but also on the position of granules are located in the seed or tuber (Hayashi et al. 1981; Hizukuri 1986; Zobel 1988; Della Valle et al. 1998). Amylose forms amorphous regions within the starch granules, while the side chains of amylopectin form crystalline regions (Jenkins and Donald 1995). Some authors have been observed co-crystalline areas consisting of amylose chains, distinguishing four types of allomorphic structures of starch (van Soest et al. 1996; van Soest and Essers 1997). The percentage of crystalline regions is very different according to the origin of the plant and it may range from 20 to 50%.

The melting temperature (T_m) and, above all, the glass transition temperature (T_g) is very high compared to the ambient temperature due to the numerous hydrogen bonds formed between the chains of this polymer. Therefore, the degradation temperature is lower than the melting temperature (Shogren 1992), making this polymer not very plastic and therefore difficult to use to produce biobased packaging as it is both rigid and brittle. In order to improve the mechanical properties of starch, it is possible to add plasticizers such as water or polyols such as glycerol and sorbitol. These substances can lower the glass transition temperature (T_g) and the melting temperature (T_m) by transforming this polymer into a plastic material that can be easily moulded (Zhang et al. 2014). The first phase for the production of "plasticized" or "thermoplastic" starch involves the destruction of the starch granules by adding water (Van Soest and Knooren 1997) and heat or thermomechanical treatment. The addition of water and the increase in temperature above 60 °C favours a phenomenon that takes the name of starch gelatinization. During this transformation, the granules swell, the amylose solubilizes and escapes from the starch granules, which collapse, losing their structure. At the end of this process, a three-dimensional structure consists of an amylose lattice that traps the water molecules.

Starch gelatinization occurs at a specific temperature, the value of which depends on the botanical origin of the starch (Genkina et al. 2007). During the storage of gelatinized starch, the reverse process takes place, which involves slow recrystallization of the amylose, which takes the name of starch retrogradation. This phenomenon involves modifying the mechanical properties of the plasticized starch, which becomes hard again and not very deformable (Van Soest and Borger 1997; Averous 2000). Temperature and relative humidity of the atmosphere where is storage material influenced the speed of this phenomenon. The addition of non-volatile plasticizing substances (polyalcohols) allows the slowing down of the retrogradation of thermo-plasticized starches (Ollett et al. 1991; Shogren and Swanson 1992; Kalichevsky and Blanshard 1993; Lourdin et al. 1995; Hulleman et al. 1999; Gaudin et al. 2000). Also, the quantity of plasticizer should be well defined since the addition of insufficient quantities cannot slow down the phenomenon of starch retrogradation (Appelqvist et al. 1993). However, these compounds are very hydrophilic and if the material is not stored at temperatures above the glass transition temperature, the retrogradation rate will increase (Lu et al. 1997; Thiewes and Steeneken 1997).

3.2.1.2   Chitin

Chitin is a polysaccharide and it is the main component of the exoskeleton of arthropods and the cell walls of yeasts and fungi (Rinaudo 2006). It has a crystalline structure and is made up of microfibrils. Chitin is constituted by N-acetyl-D-glucosamine units linked among them though β $(1 \rightarrow 4)$ link. Chitosan is obtained by deacetylation from chitin and is characterized by the degree of acetylation and the molecular weight that influenced, in turn, its viscosity and solubility (Younes and Rinaudo 2015). Chitosan can be added with glycerol to obtain a thermoplastic material similar to modified starches (Epure et al. 2011). The obtained material represents an exciting application to product precursors useful for the realization of biodegradable food packaging.

3.2.1.3   Pectin

Heteropolysaccharide formed by the polymerization of D-galacturonic acid molecules linked together by α $(1–4)$ link. Most of the carboxyl groups are esterified with methanol. The ratio between the percentage of esterified groups with methanol and the total carboxylic groups is called the degree of esterification (DE). Along the polygalacturonic-based chain, rhamnose residues are intercalated in varying proportions. These residues are concentrated in specific sites of the chain-forming branched polymers whose side chains contain rhamnose, D-galactose, L-arabinose and D-xylose (Thakur et al. 1997). Pectins are abundant in the fruit cell wall and are hydrolyzed during ripening by specific enzymes such as pectase and pectinase. The pectins cement the space between one cell and another, keeping them together and giving crunchiness to the fruit. As it ripens, this bond melts and the fruit loses consistency.

Pectins are used in the food industry as gelling agents and their behaviour depends on the amount of esterified galacturonic acid and as a function of that, they are classified as high or low ester pectins (May 1990).

3.2.1.4   Proteins

Proteins are macromolecules made up of twenty types of amino acids that differ in chemical structure and molecular weight. Generally, proteins are characterized by a complex three-dimensional structure (Gómez-Estaca et al. 2016). The main chain, in which the peptide bond links amino acids, represents the primary structure. The secondary structure consists of the spatial conformation of the chains, which can assume a spiral (alpha helix) or planar (β sheet) conformation. The tertiary structure represents the complete three-dimensional configuration that the polypeptide chain assumes in the environment in which it is located. Much of the tertiary structures can be classified as globular or fibrous. Finally, the quaternary structure derives from two or more polypeptide units linked together by weak bonds.

The proteins can be of vegetable, animal or bacterial origin and represent interesting biopolymers to realize biodegradable packaging. Among the proteins studied as precursors of biobased material derived from vegetables, there is zein extracted from corn, gluten extracted from wheat, while among those of animal origin, it is possible to include caseinates, collagen and keratin.

3.2.2 Biodegradable Polyesters

As previously mentioned, the biopolymers are biodegradable as microorganisms quickly metabolize them, unlike the petroleum-derived polymers that constitute the conventional plastics currently produced (Scholz and Khemani 2006). However, the exclusive use of plants for the extraction of these compounds would lead to high production and, therefore, cultivation of crops destined for this use, considerably aggravating the already precarious condition of the rainforests area that has already been the subject of deforestation and destruction at the order to recover land for agricultural cultivation (Goodship and Ogar 2005).

Biodegradable materials can be classified according to the history of their development, production and marketing.

The first application of biodegradable materials involved producing shopping bags made of synthetic polymers such as low-density polyethylene (LDPE) containing a small amount (5–15%) of starch and additives with pro and autoxidizing action. However, the main limitation of these materials is their disintegration into non-biodegradable fragments. Therefore, their commercial diffusion has negatively impacted consumers' opinions, contributing to reducing the environmental impact (Chiellini 2008).

An evolution of LDPE has led to the addition of pre-gelatinized starch to this polymer in a significant percentage (40–70%). In addition, this new material led to a reduction in degradation times of up to 2–3 years. The most recent innovations have led to totally biodegradable materials produced with biopolymers, classified according to their origin and production methods.

The third generation of the material consists entirely of biomaterials which can be grouped into three main categories according to their origin and production methods: extraction from biomass, chemical synthesis and production from natural o genetically modified microorganism (Chiellini 2008).

The materials most used to produce biodegradable packaging are made up of biopolymers extracted from plants or animals or produced by microorganisms. These biopolymers can be used individually or mixed with synthetic polyesters such as polylactic acid (PLA).

Starch, which is the biopolymer most present in nature and the cheapest in production terms, can be used after improving its mechanical properties by adding natural or synthetic plasticizers and thermomechanical treatments such as extrusion-cooking (Averous and Baqquillon 2004; Bastioli et al. 2020). Furthermore, it was possible to make starch-based films thanks to the addition of polyvinyl chloride (PVC) (Adeodato Vieira et al. 2011). These films can be used to produce envelopes, bags, rigid packages such as trays and thermoformed trays, thus replacing the packages produced with polystyrene and polyethylene (Ivanković et al. 2017).

Chitosan, extracted from chitin after deacetylation (Clarinval and Halleux 2005), is used to produce edible films capable of extending the shelf life of fruit and vegetables (Zhao and McDaniel 2005). This biopolymer has an antimicrobial and antifungal action that can protect food from microbiological degradation, despite its limited mechanical properties and its solubility in water (Chiellini 2008). Furthermore, the coupling of chitosan and starch for the production of multilayer films have improved the water vapour barrier and mechanical properties of these biodegradable packaging.

Vegetable proteins extracted from legumes (for example, chickpeas, peas and soybeans) or cereals (for example, wheat) or seeds of oil plants such as pistachios and sunflower are those most used for the potential production of biodegradable packaging (Dean and Yu 2005). On the other hand, albumins, caseinates, fibrinogen, silk and elastins represent the main proteins of animal origin used to produce biodegradable packaging. However, the use of these biopolymers is limited due to the difficulty to melt without a decompression treatment. In addition, they do not mix easily with other polymers due to their chemical incompatibility and the treatments to which they must be subjected to be used are costly (Ivanković et al. 2017). Other drawbacks related to the use of proteins for this particular application are their hygroscopicity and poor mechanical properties that could be improved with lipids such as stearic acid (Lodha and Netravali 2005) or the addition of glycerol, rubber gellan or k-carrageenan (Morhabed and Mittal 2007). Currently, the only film that has found a wide application is based on collagen, which could also be used in the future as an edible film (Guilbert and Cuq 2020).

References

Adeodato Vieira MG, Altenhofen da Silva MO, dos Santos LO, Beppu MM (2011) Natural-based plasticizers and bio—polymer films: a review. Eur Polym J 47:254–263

Appelqvist IAM, Cooke D, Gidley MJ, Lane SJ (1993) Thermal properties of polysaccharides at low moisture: 1—an endothermic melting process and water-carbohydrate interactions. Carbohydr Polym 20:291–299. https://doi.org/10.1016/0144-8617(93)90102-A

Averous L (2000) Properties of thermoplastic blends: starch–polycaprolactone. Polymer (guildf) 41:4157–4167. https://doi.org/10.1016/S0032-3861(99)00636-9

Averous L, Baqquillon N (2004) Biocomposite based on plasticizes starch: thermal and mechanical behaviours. Carbohydr Polym 56:111–122

Avérous L, Pollet E (2012) Biodegradable polymers. In: Green energy and technology. pp 13–39

Bastioli C, Magistrali P, Garcia SG (2020) 8. Starch-based technology. In: Handbook of Biodegradable Polymers. De Gruyter, pp 217–244

Chiellini E (2008) Environmentally compatible food packaging. Woodhead Publishing Limited

Clarinval A-M, Halleux J (2005) Classification of biodegradable polymers. In: Biodegradable polymers for industrial applications. Elsevier, pp 3–31

Cutter CN (2006) Opportunities for bio-based packaging technologies to improve the quality and safety of fresh and further processed muscle foods. Meat Sci 74:131–142. https://doi.org/10.1016/j.meatsci.2006.04.023

Davidovic A, Savic A (2010) Microbial production of bio—degradable polymer. Tehnol Acta 3:3–13

Dean K, Yu L (2005) Biodegradable protein-nanoparticle composites. In: Biodegradable polymers for industrial applications. Elsevier, pp 289–309

Della Valle G, Buleon A, Carreau PJ et al (1998) Relationship between structure and viscoelastic behavior of plasticized starch. J Rheol (n Y N Y) 42:507–525. https://doi.org/10.1122/1.550900

Denavi G, Tapia-Blácido DR, Añón MC et al (2009) Effects of drying conditions on some physical properties of soy protein films. J Food Eng. https://doi.org/10.1016/j.jfoodeng.2008.07.001

Di Mauro A, Arena E, Fallico B et al (2002) Recovery of anthocyanins from pulp wash of pigmented oranges by concentration on resins. J Agric Food Chem 50:5968–5974. https://doi.org/10.1021/jf025645s

Epure V, Griffon M, Pollet E, Avérous L (2011) Structure and properties of glycerol-plasticized chitosan obtained by mechanical kneading. Carbohydr Polym 83:947–952. https://doi.org/10.1016/j.carbpol.2010.09.003

European Bioplastics (2020) Bioplastics market data 2019. Eur Bioplastics, Berlin, Ger 1–4

Gaudin S, Lourdin D, Forssell PM, Colonna P (2000) Antiplasticization and oxygen permeability of starch-sorbitol films. Carbohydr Polym. https://doi.org/10.1016/S0144-8617(99)00206-4

Genkina NK, Wikman J, Bertoft E, Yuryev VP (2007) Effects of structural imperfection on gelatinization characteristics of amylopectin starches with A- and B-type crystallinity. Biomacromol 8:2329–2335. https://doi.org/10.1021/bm070349f

Geueke B, Groh K, Muncke J (2018) Food packaging in the circular economy: Overview of chemical safety aspects for commonly used materials. J Clean Prod 193:491–505

Gómez-Estaca J, Gavara R, Catalá R, Hernández-Muñoz P (2016) The potential of proteins for producing food packaging materials: a review. Packag Technol Sci 29:203–224. https://doi.org/10.1002/pts.2198

Goodship V, Ogar E (2005) Polymer processing with supercritical fluids. Polim Časopis Za Plast i Gumu 26:47–48

Guilbert S, Cuq B (2020) 11. Material formed from proteins. In: Handbook of Biodegradable Polymers. De Gruyter, pp 299–338

Hayashi A, Kinoshita K, Miyake Y (1981) The conformation of amylose in solution. I. Polym J 13:537–541. https://doi.org/10.1295/polymj.13.537

Hizukuri S (1986) Polymodal distribution of the chain lengths of amylopectins, and its significance. Carbohydr Res 147:342–347. https://doi.org/10.1016/S0008-6215(00)90643-8

Hulleman SH, Kalisvaart M, Janssen FH et al (1999) Origins of B-type crystallinity in glycerol-plasticised, compression-moulded potato starches. Carbohydr Polym 39:351–360. https://doi.org/10.1016/S0144-8617(99)00024-7

Imre B, García L, Puglia D, Vilaplana F (2019) Reactive compatibilization of plant polysaccharides and biobased polymers: Reviewon current strategies, expectations and reality. Carbohydr Polym 209:20–37; https://doi.org/10.1016/j.carbpol.2018.12.082

Ivanković A, Zeljko K, Talić S, Lasić M (2017) Biodegradable packaging in food industry. J Food Saf Food Qual 68:23–52. https://doi.org/10.2376/0003-925X-68-26

Jenkins PJ, Donald AM (1995) The influence of amylose on starch granule structure. Int J Biol Macromol 17:315–321. https://doi.org/10.1016/0141-8130(96)81838-1

Kalichevsky MT, Blanshard JMV (1993) The effect of fructose and water on the glass transition of amylopectin. Carbohydr Polym 20:107–113. https://doi.org/10.1016/0144-8617(93)90085-I

Laufenberg G, Kunz B, Nystroem M (2003) Transformation of vegetable waste into value added products: Bioresour Technol 87:167–198. https://doi.org/10.1016/S0960-8524(02)00167-0

Lodha P, Netravali AN (2005) Thermal and mechanical properties of environment-friendly 'green' plastics from stearic acid modified-soy protein isolate. Ind Crops Prod 21:49–64. https://doi.org/10.1016/j.indcrop.2003.12.006

Lourdin D, Della VG, Colonna P (1995) Influence of amylose content on starch films and foams. Carbohydr Polym 27:261–270. https://doi.org/10.1016/0144-8617(95)00071-2

Lu T, Jane J-I, Keeling PL (1997) Temperature effect on retrogradation rate and crystalline structure of amylose. Carbohydr Polym 33:19–26. https://doi.org/10.1016/S0144-8617(97)00038-6

May CD (1990) Industrial pectins: sources, production and applications. Carbohydr Polym 12:79–99. https://doi.org/10.1016/0144-8617(90)90105-2

Morhabed E, Mittal GS (2007) Formulation and process conditions for biodegradable/edible soy-based packaging trays. Packg Technol Sci 20:1–15

Ollett AL, Parker R, Smith AC (1991) Deformation and fracture behaviour of wheat starch plasticized with glucose and water. J Mater Sci 26:1351–1356. https://doi.org/10.1007/BF00544476

Ötles SÖS (2004) Biobased packaging materials for the food industry—types of biobased packaging materials. J Oil Soap Cosmet 53:116–119

Prieto A (2016) To be, or not to be biodegradable… that is the question for the bio-based plastics. Microb Biotechnol 9:652–7. https://doi.org/10.1111/1751-7915.12393

Rinaudo M (2006) Chitin and chitosan: properties and applications. Prog Polym Sci 31:603–632

Salgado PR, Schmidt VC, Molina Ortiz SE et al (2008) Biodegradable foams based on cassava starch, sunflower proteins and cellulose fibers obtained by a baking process. J Food Eng 85:435–443. https://doi.org/10.1016/j.jfoodeng.2007.08.005

Scholz C, Khemani K (2006) Degradable polymers and materials. American Chemical Society, Washington, DC

Shogren RL (1992) Effect of moisture content on the melting and subsequent physical aging of cornstarch. Carbohydr Polym 19:83–90. https://doi.org/10.1016/0144-8617(92)90117-9

Shogren RL, Swanson CLTA (1992) Extrudates of cornstarch with urea and glycols: structure/mechanical property relations. Starch—Stärke 44:335–338

Siracusa V, Rocculi P, Romani S, Rosa MD (2008) Biodegradable polymers for food packaging: a review. Trends Food Sci Technol 19:634–643. https://doi.org/10.1016/j.tifs.2008.07.003

Suyatma NE, Copinet A, Tighzert L, Coma V (2004) Mechanical and barrier properties of biodegradable films made from chitosan and poly (lactic acid) blends. J Polym Environ 12:1–6. https://doi.org/10.1023/B:JOOE.0000003121.12800.4e

Thakur BR, Singh RK, Handa AK, Rao MA (1997) Chemistry and uses of pectin—a review. Crit Rev Food Sci Nutr 37:47–73. https://doi.org/10.1080/10408399709527767

Thiewes HJ, Steeneken PAM (1997) The glass transition and the sub-Tg endotherm of amorphous and native potato starch at low moisture content. Carbohydr Polym 32:123–130. https://doi.org/10.1016/S0144-8617(96)00133-6

Van de Velde K, Kiekens P (2002) Biopolymers: overview of several properties and consequences on their applications. Polym Test 21:433–442. https://doi.org/10.1016/S0142-9418(01)00107-6

van Soest JJG, Essers P (1997) Influence of amylose-amylopectin ratio on properties of extruded starch plastic sheets. J Macromol Sci Part A 34:1665–1689. https://doi.org/10.1080/10601329708010034

van Soest JJG, Hulleman SHD, de Wit D, Vliegenthart JFG (1996) Crystallinity in starch bioplastics. Ind Crops Prod 5:11–22. https://doi.org/10.1016/0926-6690(95)00048-8

Van Soest JJG, Borger DB (1997) Structure and properties of compression-molded thermoplastic starch materials from normal and high-amylose maize starches. J Appl Polym Sci 64:631–644. https://doi.org/10.1002/(SICI)1097-4628(19970425)64:4%3c631::AID-APP2%3e3.0.CO;2-O

Van Soest JJG, Knooren N (1997) Influence of glycerol and water content on the structure and properties of extruded starch plastic sheets during aging. J Appl Polym Sci 64:1411–1422. https://doi.org/10.1002/(SICI)1097-4628(19970516)64:7%3c1411::AID-APP21%3e3.0.CO;2-Y

Weber CJ, Haugaard V, Festersen R, Bertelsen G (2002) Production and applications of biobased packaging materials for the food industry. Food Addit Contam 19:172–177. https://doi.org/10.1080/02652030110087483

Younes I, Rinaudo M (2015) Chitin and chitosan preparation from marine sources. Structure properties and applications. Mar Drugs 13:1133–1174. https://doi.org/10.3390/md13031133

Zhang Y, Rempel C, McLaren D (2014) Thermoplastic starch. In: Innovations in food packaging. Elsevier, pp 391–412

Zhao Y, McDaniel M (2005) Sensory quality of foods associated with edible film and coating systems and shelf-life extension. In: Innovations in food packaging. Elsevier, pp 434–453

Zobel HF (1988) Molecules to granules: a comprehensive starch review. Starch—Stärke 40:44–50. https://doi.org/10.1002/star.19880400203

Chapter 4
Environmental Impact Generated by Technologies and by Processes Used to Produce Biodegradable Food Packaging

Abstract Many are the cheap vegetable sources (grape, tomato, pineapple, citrus fruits, cereal, etc.) derived from agricultural by-products. The carbon-rich precursors are used to produce bio-based polymers through microbial, biopolymer blending and chemical methods. Even though the enhancement of advanced synthetic methods and the application of biofilms in smart/intelligent food packaging, commercial production is limited by cost, the economics of production, useful life, biodegradation concerns and adequate availability of agricultural wastes. The innovative and production cost of technological processing is critical to facilitating bio-based polymers' commercial production and replacing synthetic polymers. This chapter gives an overview of bio-based food packaging's leading technological processing production, highlighting their advantages and limits and the potential applications in food sectors.

Keywords Agricultural waste · Chemical extraction · Extrusion-cooking processing · Gas permeability · Microbial processing polymers · Sustainability

Abbreviations

AD	Anaerobic digester-based compost
LDPE	Low density polyethylene
ME	Mint plant extracts
OWC	Olive waste-based compost
PHBV	Poly(3-hydroxybutyrateco-3-hydroxyvalerate)
PHA	Polyhydroxyalkanoates
PHB	Polyhydroxybutyrate
PLA	Poly-lactic acid
PP	Polypropylene
PVA	Polyvinyl alcohol
PE	Pomegranate peel extract
PLE	Pressurized liquid extraction

SE   Solvent extraction
TEOS   Tetraethoxysilane
TPS   Thermoplastic starch
WVTR   Water Vapour Transmission Rate

4.1 Technological Processing Used to Produce Eco-Friendly Food Packaging

The critical factors of biodegradable packaging production depend on the finding and choice of raw materials, the suitable production technologies and the identification of business strategies that reduce production costs without compromising the quality of the products (Rosentrater and Otieno 2006). The areas involved in the design of sustainable packaging are numerous and very different. They concern the industrial design, the environmental sciences and marketing, even if the goal and the fundamental aspect that must be considered concerning the disposal procedure of this material (Han 2005).

All technologies used to produce biodegradable packaging derived from by-products or waste from agriculture and the food industry involve the destruction of the original molecular structure (intermolecular non-covalent and covalent bonds), the arrangement and orientation of the free polymer chains, according to the desired shape and structure and the formation and the stabilization of a new three-dimensional polymer structure, after the elimination of the breaking agent (Guilbert et al. 2002).

The production processes of biodegradable packaging are based on two technologies (Fig. 4.1):

1. polymer dispersion or solubilization into various solvents as a function of its chemical characteristics;
2. melting of polymers through thermal treatment and modelling by mechanical processing like compression, cutting and shaping.

The polymers dispersion/solubilization phase is followed by applying the obtained mixture through casting, spraying or dipping techniques on a suitable rigid support and drying.

The thermoforming technology favours the biopolymers' glass transition to obtain a plastic mass that can ease modelled. After the shaping phase, the polymeric structure is stabilized by cooling.

The transformation of biopolymers by thermo-fusion to produce biodegradable packaging can be carried out with the techniques used for plastics through appropriate modifications and optimization of the process parameters (Thunwall et al. 2008; Cruz-Romero and Kerry 2008). The main techniques used to produce thermoplastic materials include a thermo-mechanical treatment by extrusion and

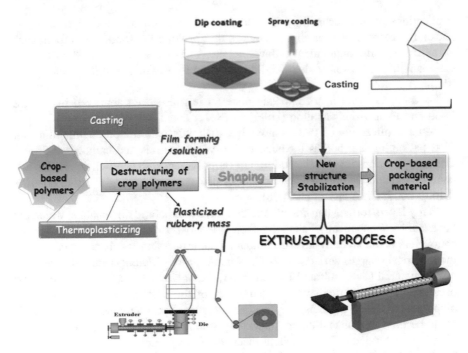

Fig. 4.1 Schematic representation of the two technological processes used to form crop-based packaging materials

moulding by injection and compression (Bastioli and Bettarini 2020). Extrusion allows obtaining films (thickness less than 100 microns) from thermoplastic material using two techniques: extruded cast film and extruded blow film (Han et al. 2018). An example of food packaging produced through the thermoforming of biopolymers derived from agricultural by-products is represented by the use of PLA (Garlotta 2001) and, in particular, by the production of PLA bottles devoted to packaging mineral water and milk (Cruz-Romero and Kerry 2008). Other applications concern the production of trays and containers for meat products or frozen foods or food heating by microwaves, etc. (Prieto 2016).

Nanotechnology represents one of the most recent and promising technologies to realize biodegradable packaging. Nanocomposite materials are very light and have a low environmental impact than conventional plastic food packaging (Youssef and El-Sayed 2018). In addition, these innovative materials may improve the mechanical and barrier properties of thermoplastic starch-based food packaging (Stollman et al. 2000).

Nanotechnology enables the design, formation and manipulation of structures and materials with a size of a few nanometers (1–100 nm). The application of this new science is extensive and varied, representing a considerable potential for the technological and economic development of future generations (Sozer and Kokini 2009; Ravichandran 2009). For example, one of the possible innovative applications of nanotechnology in food packaging could be the design of active packaging capable of releasing substances that prolong the shelf life of food, such as antimicrobials, antioxidants, flavours, etc. (Vermeiren et al. 1999; Cha and Chinnan 2004; Lagaron et al. 2005; Sinharay and Bousmina 2005; Kuswandi 2017; Mellinas et al. 2020).

Food contact with nanocomposite material has also been approved by the US Food and Drug Administration (USFDA) (Sozer and Kokini 2009).

Some applications of these innovative materials concern producing films for food packaging and bottles for beer, edible vegetable oils and carbonated drinks. These applications involve the addition of biopolymers with nano-clay particles that confer a remarkable improvement of barrier properties of packaging up to 80–90% (Brody 2007; Chaudhry et al. 2008).

Many layers formed a nanolaminate that have nanometric dimensions, which are physically and chemically linked together. The most performing method to link the nanolayers is the layer-by-layer technique. This processing involves the coating of the electrical charges surfaces with films made up of different layers of nanocomposite material (Schlenoff and Decher 2003). Nanolaminates could produce edible packaging for dairy products or other types of food (Weiss et al. 2006). Many biopolymers such as proteins, polysaccharides, polar lipids, emulsifiers and colloids could be used to realize different nanocomposite layers to obtain edible films with high barrier properties and excellent mechanical properties.

The materials containing nano-silicates consist of two layers having a thickness of 1 nm and a length of several microns, as a function of used silicate particles. The addition of these nanocomposite materials in the polymer matrices entangle the permeation pathway of gas molecules. This change of packaging material increases the tortuosity of the inter and intramolecular gaps improving barrier properties (Bharadwaj et al. 2002; Cabedo et al. 2004; Mirzadeh and Kokabi 2007).

4.2   Primary Applications of Biodegradable Packaging Made up of Agri-Food By-Products and Waste

The keystone of biopolymers' choice to produce food packaging is based on the mechanical properties of the materials obtained from them. The tensile strength, compression and yield strength represent the essential characteristics to realize materials devoted to food packaging production. These properties are greatly influenced by the extraction method of biopolymers. Therefore, materials consisting of fibres such as cellulose can confer good mechanical resistance even if it is necessary to use polymers not excessively complex with many substituent groups, such as the cellulose precursor, since they may not be biodegradable (Vroman and Tighzert 2009). Agricultural waste comprises post-harvest waste, by-products of food processing such as coconut shells (Nunes et al. 2020), potato peels (Xie et al. 2020), fruit peels (Bashir et al. 2018), and fruit seeds (Santana et al. 2018), which generally derived from production activities of farms and/or food industries. Agricultural by-products are the primary source of bio-based plastics, plasticizers and antioxidant additives (Xie et al. 2020). In addition, vegetable-based agricultural waste is a precious source of polysaccharides (Di Donato et al. 2020).

The use of agricultural or food industry by-products and waste as fertilizers or compost has a considerable negative effect on global warming because of the high production of carbon dioxide (Diacono et al. 2019). Therefore, the production of biopolymers by agricultural or food industry by-products and waste to produce bio-based materials represents a valid alternative to plastic (produced from a non-renewable and non-biodegradable source) and also a winning strategy to reduce carbon dioxide production. Furthermore, these applications for the reuse and recovery of waste material are innovative and have a reduced impact on the environment, as shown by the life cycle assessment analysis (LCA).

The main problem to replace plastic with bio-based materials is the low productivity and the poor mechanical and barrier properties of biodegradable packaging. In fact, in 2018, only 2.1 million tons of biopolymers were produced compared to 7 million tons required in 2020 to replace 46% of global plastic production that correspond to 400 million tons (DeGruson 2016; Vox et al. 2016; Gontard et al. 2018). The difficulty of accurately estimating the future production of biodegradable materials is due to the continuous development of innovative production technologies and the complexity of the supply chain and the production process (Ramesh Kumar et al. 2020).

In order to increase the availability and consequent productivity of biopolymers extracted from agricultural by-products, it would be appropriate and necessary to concentrate the production sites in specific geographical areas according to the availability of agricultural waste. For example, microalgae are abundant in coastal areas, while in tropical areas and Asia (India and China), significant quantities of by-products deriving from fruit and vegetable cultivation are produced (Mathiot et al. 2019; Nunes et al. 2020; Sharma et al. 2020).

The criteria to choose agricultural by-products and waste to produce biopolymers are related to the chemical composition (content in starch, cellulose, hemicellulose, lignin); the availability and impact on agricultural supply chains and food safety; the extraction or synthesis processes; the desired properties of the end product and the biodegradability (Mose and Maranga 2011). Table 4.1 shows the chemical sample of some agricultural by-products. It is possible to observe the high percentage of cellulose in the corn stalks. The high cellulose content increases the material's mechanical strength but reduces its biodegradability since microorganisms slowly metabolize this biopolymer due to its complex chemical structure (Iwata 2015; Maraveas 2019). Therefore, the right compromise must be found when a material rich in cellulose was chosen since it is necessary to guarantee high biodegradability and good mechanical strength.

Starch is another fundamental biopolymer to produce biodegradable materials since it influences the thickness of the film. A high presence of starch in agricultural by-products and waste allows the production of films with an optimal thickness (about 0.099–0.1599 mm) (Santana et al. 2018).

Table 4.1 Chemical composition of common forms of agricultural waste (Sadh et al. 2018)

Agro-industrial wastes	Chemical composition (% w/w)					
	Cellulose	Hemicellulose	Lignin	Ash (%)	Total solids (%)	Moisture (%)
Sugarcane bagasse	30.2	56.7	13.4	1.9	91.66	4.8
Rice straw	39.2	23.5	36.1	12.4	98.62	6.58
Corn stalks	61.2	19.3	6.9	10.8	97.78	6.40
Sawdust	45.1	28.1	24.2	1.2	98.54	1.12
Sugar beet waste	26.3	18.5	2.5	8	87.5	12.4
Barley straw	33.8	21.9	13.8	11	–	-
Cotton stalks	58.5	14.4	21.5	9.98	–	7.45
Oat straw	39.4	27.1	17.5	8	–	–
Soya stalks	34.5	24.8	19.8	10.39	–	11.84
Sunflower stalks	42.1	29.7	13.4	11.17	–	–
Wheat straw	32.9	24.0	8.9	6.7	95.6	7

In addition to the amount, it is crucial to consider the composition and starch structure when this biopolymer may use to realize biodegradable packaging. In fact, the amounts of amylose and amylopectin change according to the nature of the source by which it is extracted. For example, the starch of microalgae is very different from that present in jackfruits. Therefore, the two types of starch must be submitted to the different production processes and the industrial applications (Mose and Maranga 2011; Santana et al. 2018; Mathiot et al. 2019).

Different natural or synthetic polymers (tetraethoxysilane—TEOS, polyvinyl alcohol—PVA and chitosan) have been added to thermoplastic starch to improve its barrier and mechanical properties (Treinyte et al. 2018). In particular, chitosan improves and promotes starch binding with synthetic polymers such as TEOS and PVA (Farzadnia et al. 2018). However, the chemicals used to promote cross-linking of biopolymers such as formaldehyde are toxic and non-biodegradable; their use must be limited (Bashir et al. 2018).

The use of starch as a raw material to produce bio-based materials on a commercial scale is limited because the crops rich in this biopolymer represent the primary nutritional source of many populations, especially the poorest ones.

Some studies have highlighted the possibility to obtain biopolymers by skins of some fruits such as pineapples or vegetables such as tomatoes to improve the mechanical properties of the obtained-based materials (Vega-Castro et al. 2016; Heredia-Guerrero et al. 2019). These applications are carried out using suitable extraction systems and the optimization requires a treatment time of 60 h and a pH value of 9 (Vega-Castro et al. 2016). Biopolymers produced by fermentation of vegetal source, the environment acidification with sulfuric acid or dipotassium phosphate or ammonium sulfate can improve extraction yield. However, the massive use of these substances may cause excessive acidification and eutrophication of the environment by increasing environmental impact (Yates and Barlow 2013).

Patil et al. (2018) observed that the mechanical properties of natural fibres and epoxy resins are improved by adding 10–30% of lemon peel and sweet lime powder. The positive action of the dehydrated peel of these fruits is due to high quantities (about 90%) of cellulose, lignin and crude fibres. Nevertheless, this application is complicated to extend on a large scale because of the limited availability of these by-products. In fact, lemon and sweet lime skin are edible or used to extract bioactive compounds such as polyphenols.

Biopolymers can also be produced through fermentation processes by Gram-negative and Gram-positive microorganisms that can metabolize agricultural by-products (source of carbon). An example of a biopolymer produced by microbial synthesis is polyhydroxyalkanoates (PHA). The factors that influence the yield in the microbiological synthesis of biopolymers produced by agricultural waste are the pH, the availability of essential nutrients such as carbon, phosphorus and nitrogen, the composition and type of substrate necessary to metabolism microorganisms (Anjum et al. 2016; Tsang et al. 2019). Polyhydroxybutyrate (PHB) has good mechanical properties comparable to that of plastic polymers such as polypropylene (Khardenavis et al. 2007). However, the limiting factor for a broad-spectrum application is the high production cost (nine times higher than other bio-based materials). In the last years, the PHB has become cheaper by using it as a carbon source by-products of rice and jowar processing (Khardenavis et al. 2007).

De Pilli et al. (2016) proposed a biodegradable and compostable material devoted to producing new food packaging. The innovative elements concern using the complete waste derived from food industries without the extraction of biopolymers. The absence of the extraction phase involves no use of organic solvents that determines the simplification of the process and low production costs.

Remarkable are the advantages that derive from this type of product, like the solution of disposal problem of waste derived from food industries that, in some cases, represent a considerable impact on production cost. This new production of biodegradable packaging can reduce, then at zero, the disposal cost relating to this type of waste and confer an added value of these materials because they become the raw material in the production of the food packaging. Furthermore, for some industries, the life cycle of waste materials may be carried out in the same production site, eliminating transport and storage costs. Other advantages concern the decrease of cost relating to the purchase of packaging material. Moreover, the replacement of packaging made from plastic materials; no readily biodegradable, with those fully biodegradable, reduces the environmental impact on the production company itself. This feature is required both by consumers and by the new European policies (HORIZON 2020). Consequently, the industries that will use this packaging material could gain market share and some incentives for companies that reduce environmental pollution through more sustainable production.

The simplicity of the production process of the material patented by De Pilli et al. (2016) is highlighted in Fig. 4.2. The flow chart reported in Fig. 4.2 is an example of applying asparagus waste to produce biodegradable and compostable packaging. This production process consists of two preliminary drying and grinding operations, which can transform the waste into a material that can be processed by extrusion-cooking. First, the biopolymers present in asparagus waste and cereal flours (added to the formula) are modified by heat treatment (maximum temperature reached 120 °C) and mechanical action (high shear stresses). Then, the mechanical properties are improved with the addition of material rich in lipids or polyols with a natural origin. After, the extruded material is shaped by press mould and, finally, it is dehydrated.

The material patented by De Pilli et al. (2016) showed excellent mechanical properties without the use of complex synthetic plasticizers chemical-physical treatments. In fact, the flexibility values were similar to those of the primary conventional materials (cardboard, polypropylene, aluminium and polystyrene) used to produce food packaging and, the breaking load values were comparable to that of aluminium containers (Fig. 4.3).

The effects of low temperatures on mechanical properties change of the material based on asparagus waste were evaluated, storing the material for eight days at different temperatures: −20 °C; +4 °C; and +25 °C (control sample). On the ninth day, all samples were exposed without any cover at room temperature for 24. After that, the flexibility and the breaking load were evaluated. The results showed that the samples stored at different temperatures had similar flexibility and tensile strength values. Therefore, this material could be used to produce packaging devoted to chilled and frozen food and beverages.

Furthermore, biodegradability and compostability tests have shown that the asparagus waste-based material is $97 + 0.84\%$ biodegradable and, according to the parameters reported by the UNI EN ISO 13432: 2002 standard and by the Italian Legislative Decree no. 75/2010, also resulted compostable (Table 4.2). Therefore, compostability offers an added value to this material which not only can be disposed of as organic waste but can also be used as an organic fertilizer in the final phase of its life cycle.

Another exciting aspect of the material produced with asparagus waste concerns the barrier properties against water vapour. In fact, the results relating to the water vapour transmission rate (WVTR) showed that this material has a permeability lower than cellophane (Table 4.3).

Figure 4.4 shows some objects produced with different types of waste from food industries through the production technology patented by De Pilli et al. (2016). These results highlight the enormous potential applicative of this innovation for the production of food packaging.

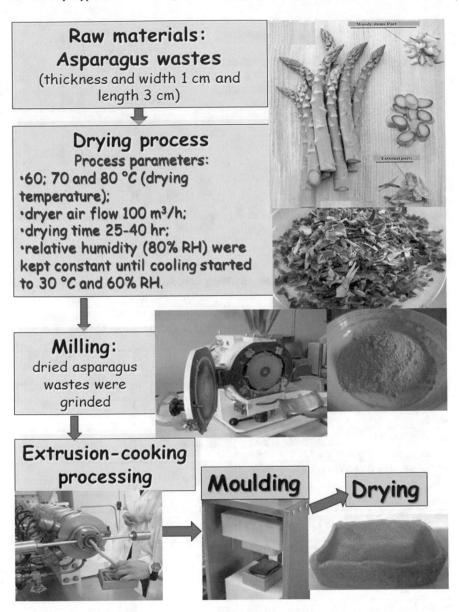

Fig. 4.2 Flow chart to produce biodegradable packaging based on asparagus wastes

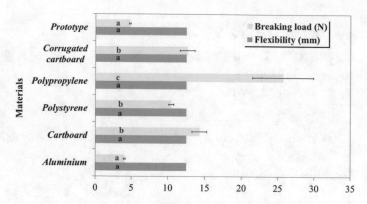

Fig. 4.3 Break load and flexibility of the biodegradable and compostable material for packaging and the main materials used for food packaging

Table 4.2 Tests of assessment of the biodegradability and compostability of the material patented by De Pilli et al. (2016)

	Mean values	St. Dev.	Limits, according to DIgs. 75/2010	Limits of the norm UNI EN ISO 13432:2002
Biodegradability (%)	97.40	0.87	*90*	*90*
pH	7.28	0.19	*6–8.5*	
Humidity at 105 °C (%)	11.90	0.92	*<50*	
Total solids, fixed total solids	2.24	0.22	–	
Residue 550 °C (%)	98.20	0.46	–	
*Total organic carbon (%)	7.85	1.66	*20*	
*Total nitrogen (as N) (%)	1.77	0.18	–	
C/N ratio	32.85	2.32	*25*	
*Organic nitrogen	1.62	0.13	–	
Organic N/total N ratio (%)		91.85	2.64	*80*
*Plastic materials, glass and metals ≥ 2 mm(%)	0.00	0.00	*0.5*	*10*
*Lithoid interts ≥ 5 mm (%)	0.00	0.00	*5*	
*Cadmium mg/kg	0.03	0.02	*1.5*	*0.5*
*Hexavalent chromium mg/kg	<0.01	0.00	*0.5*	*50*
*Mercury mg/kg	<0.01	0.00	*1.5*	*0.5*
*Nickel mg/kg	0.96	0.10	*100*	*25*
*Lead mg/kg	1.01	0.13	*140*	*50*
*Copper mg/kg	2.21	0.41	*2.03*	*50*
*Zinc mg/kg	13.54	1.91	*11.87*	*150*

(continued)

Table 4.2   (continued)

	Mean values	St. Dev.	Limits, according to DIgs. 75/2010	Limits of the norm UNI EN ISO 13432:2002
Salmonella	Absent/ 25 g		*Absent/25 g*	
Escherichia coli	0.00	0.00	*0*	
Static breathing index (as O2)	173.67	4.51	*178*	
Germination index	93.33	3.06	*90*	90
Nitrogen mineralisation index	1.53	0.24	*1.37*	

Table 4.3 WVTR values of the main material food packaging and biodegradable material patented by De Pilli et al. (2016)

Material	WVTR (30 °C, 90% UR) g/m^2/day
Aluminium	0
The material of the invention	$3.16 * 10^{-3}$
Cardboard	$2.00 * 10^{-2}$
Cellophane	$3.75 * 10^{-2}$

4.3   Environmental Impacts and Health Implication of Biodegradable Packaging Made up of Agri-Food By-Products and Waste. Advantages and Disadvantages

The use of biodegradable packaging produced using biopolymers deriving from agricultural by-products or waste or the food industry to replace plastic material undoubtedly has many advantages. The production processes of these materials are certainly more energy efficient than the production of plastic material. The biopolymers have a posive impact on the environment and human health and thanks to their biodegradability. Moreover, they can be disposed of very easily, quickly and at a low cost. Furthermore, the production of bipolymers results in reduced carbon dioxide emissions, thus helping to reduce global warming and climate change. Another advantage is undoubtedly using renewable sources for their pro-duction, unlike plastics that derive from a source destined to run out, such as oil (Sangsuwan et al. 2008). However, biopolymers have some limitations, such as the high amount of vegetable raw material for their synthesis. Therefore, it is necessary to innovate production systems to reduce the extensive use of agricultural land to cultivate biomass used to extract biopolymers. In addition, it is necessary to improve and optimize the production processes of biopolymers to improve the economic and sustainability aspects (Zhao and McDaniel 2005; Sangsuwan et al.

Fig. 4.4 Example of some packaging produced through the biodegradable material

2008). Another important aspect is developing effective composting systems that also require efficient collection systems of treated material (Wiles and Scott 2006).

There are few studies relating to the effect of biopolymers on human health. In any case, it has been shown that the direct and negative impacts of polymers on human health and flora and fauna are limited or absent since biopolymers decompose in compounds that do not negatively affect human health (Gross and Kalra 2002; Flint et al. 2012).

An indirect negative impact of the production of biopolymers on human health can be attributed to the production of greenhouse gases released during their synthesis (which can cause respiratory problems such as asthma) and the use of chemicals for the agricultural production of the biomass (which can favour the development of carcinomas) (Weisenburger 1993).

However, it is possible to find a considerable difference between the amount of greenhouse gas produced from plastic and biopolymer processing. Therefore, the use of bio-based materials is certainly more advantageous than petroleum derivatives. Consequently, it is possible to affirm that biodegradable packaging is an excellent alternative to plastic.

4.4 Potential Developments

Numerous factors, including political and legislative changes and global demand for foods and energy resources, will influence bio-based packaging materials' growth and success. However, there is no doubt that bio-based materials used to produce food packaging will increase since to decrease the production cost and the mechanical and barrier to gas transmission properties will be improved. In addition, the use of biodegradable packaging could be encouraged and expanded by supermarket chains by the application of discounts on food packed with biodegradable materials and through specific requests to their suppliers. Already necessary research has been undertaken to improve mechanical and barrier properties with some small commercialisation. Cost is undoubtedly a limitation to the widespread adoption of bio-based packaging materials, but as production capacity increases, costs will fall. One barrier to reducing costs is the increase in biofuels production, which in many cases are competing for the same raw materials (corn and maize) as bio-based packaging, putting upward pressure on raw material costs. From a commercial point of view, the biggest unanswered question is whether consumers will enthusiastically embrace the use of bio-based packaging materials. It is not easy to forecast consumers' future preference between bio-based packaging or conventional recycled materials or if both methods will be chosen. Indeed, future populations' main aim must be to propose to reach the sensibilization and dissemination of the social culture of respect and environmental protection by forming future generations.

References

Anjum A, Zuber M, Zia KM et al (2016) Microbial production of polyhydroxyalkanoates (PHAs) and its copolymers: a review of recent advancements. Int J Biol Macromol 89:161–174. https://doi.org/10.1016/j.ijbiomac.2016.04.069

Bashir A, Jabeen S, Gull N et al (2018) Co-concentration effect of silane with natural extract on biodegradable polymeric films for food packaging. Int J Biol Macromol. https://doi.org/10.1016/j.ijbiomac.2017.08.025

Bastioli C, Bettarini F (2020) General characteristics, processability, industrial applications and market evolution of biodegradable polymers. In: Handbook of biodegradable polymers

Bharadwaj RK, Mehrabi AR, Hamilton C et al (2002) Structure-property relationships in cross-linked polyester-clay nanocomposites. Polymer (guildf). https://doi.org/10.1016/S0032-3861(02)00187-8

Brody AL (2007) Case studies on nanotechnologies for food packaging. Food Technol 7:102–107

Cabedo L, Giménez E, Lagaron JM et al (2004) Development of EVOH-kaolinite nanocomposites. Polymer (guildf). https://doi.org/10.1016/j.polymer.2004.05.018

Cha DS, Chinnan MS (2004) Biopolymer-based antimicrobial packaging: a review. Crit Rev Food Sci Nutr. https://doi.org/10.1080/10408690490464276

Chaudhry Q, Scotter M, Blackburn J et al (2008) Applications and implications of nanotechnologies for the food sector. Food Addit Contam—Part A Chem Anal Control Expo Risk Assess. https://doi.org/10.1080/02652030701744538

Cruz-Romero M, Kerry JP (2008) Crop-based biodegradable packaging and its environmental implications. CAB Rev Perspect Agric Vet Sci Nutr Nat Resour 3. https://doi.org/10.1079/PAVSNNR20083074

De Pilli T, Derossi A, Severini C (2016) Biodegradable and compostable material for packaging obtained from the use of the whole wastes of production of food industries. 1–17

DeGruson ML (2016) Biobased polymer packaging. In: Reference module in food science

Di Donato P, Taurisano V, Poli A et al (2020) Vegetable wastes derived polysaccharides as natural eco-friendly plasticizers of sodium alginate. Carbohydr Polym. https://doi.org/10.1016/j.carbpol.2019.115427

Diacono M, Persiani A, Testani E et al (2019) Recycling agricultural wastes and by-products in organic farming: biofertilizer production, yield performance and carbon footprint analysis. Sustain. https://doi.org/10.3390/su11143824

Farzadnia N, Bahmani SH, Asadi A, Hosseini S (2018) Mechanical and microstructural properties of cement pastes with rice husk ash coated with carbon nanofibers using a natural polymer binder. Constr Build Mater. https://doi.org/10.1016/j.conbuildmat.2018.04.205

Flint S, Markle T, Thompson S, Wallace E (2012) Bisphenol A exposure, effects, and policy: a wildlife perspective. J Environ Manage. https://doi.org/10.1016/j.jenvman.2012.03.021

Garlotta D (2001) A literature review of poly(lactic acid). J Polym Environ. https://doi.org/10.1023/A:1020200822435

Gontard N, Sonesson U, Birkved M et al (2018) A research challenge vision regarding management of agricultural waste in a circular bio-based economy. Crit Rev Environ Sci Technol. https://doi.org/10.1080/10643389.2018.1471957

Gross RA, Kalra B (2002) Biodegradable polymers for the environment. Science 803 https://doi.org/10.1126/science.297.5582.803

Guilbert S, Gontard N, Morel MH, et al (2002) Formation and properties of wheat gluten films and coatings. In: Protein-based films and coatings

Han JW, Ruiz-Garcia L, Qian JP, Yang XT (2018) Food packaging: a comprehensive review and future trends. Compr Rev Food Sci Food Saf. https://doi.org/10.1111/1541-4337.12343

Heredia-Guerrero JA, Caputo G, Guzman-Puyol S et al (2019) Sustainable polycondensation of multifunctional fatty acids from tomato pomace agro-waste catalyzed by tin (II) 2-ethylhexanoate. Mater Today Sustain. https://doi.org/10.1016/j.mtsust.2018.12.001

Iwata T (2015) Biodegradable and bio-based polymers: future prospects of eco-friendly plastics. Angew Chemie Int Ed 54:3210–3215. https://doi.org/10.1002/anie.201410770

Khardenavis A, Sureshkumar M, Mudliar S, Chakrabarti T (2007) Biotechnological conversion of agro-industrial wastewaters into biodegradable plastic, poly β-hydroxybutyrate. Bioresour Technol 98:3579–3584. https://doi.org/10.1016/j.biortech.2006.11.024

Kuswandi B (2017) Environmental friendly food nano-packaging. Environ Chem Lett 15:205–221. https://doi.org/10.1007/s10311-017-0613-7

Lagaron JM, Cabedo L, Cava D et al (2005) Improving packaged food quality and safety. Part 2: nanocomposites. Food Addit Contam 22:994–998. https://doi.org/10.1080/02652030500239656

Maraveas C (2019) Environmental sustainability of greenhouse covering materials. Sustainability 11:6129. https://doi.org/10.3390/su11216129

Mathiot C, Ponge P, Gallard B et al (2019) Microalgae starch-based bioplastics: screening of ten strains and plasticization of unfractionated microalgae by extrusion. Carbohydr Polym 208:142–151. https://doi.org/10.1016/j.carbpol.2018.12.057

Mellinas C, Ramos M, Jiménez A, Garrigós MC (2020) Recent trends in the use of pectin from agro-waste residues as a natural-based biopolymer for food packaging applications. Materials (basel) 13:673. https://doi.org/10.3390/ma13030673

Mirzadeh A, Kokabi M (2007) The effect of composition and draw-down ratio on morphology and oxygen permeability of polypropylene nanocomposite blown films. Eur Polym J 43:3757–3765. https://doi.org/10.1016/j.eurpolymj.2007.06.014

Mose BR, Maranga SM (2011) A review on starch based nanocomposites for bioplastic materials. Former part J Mater Sci Eng 239–245

Nunes LA, Silva MLS, Gerber JZ, Kalid de RA (2020) Waste green coconut shells: Diagnosis of the disposal and applications for use in other products. J Clean Prod 255:120169. https://doi.org/10.1016/j.jclepro.2020.120169

Patil AY, Hrishikesh NU, Basavaraj GD et al (2018) Influence of bio-degradable natural fiber embedded in polymer matrix. Mater Today Proc 5:7532–7540. https://doi.org/10.1016/j.matpr.2017.11.425

Prieto A (2016) To be, or not to be biodegradable… that is the question for the bio-based plastics. Microb Biotechnol 9:652–657. https://doi.org/10.1111/1751-7915.12393

RameshKumar S, Shaiju P, O'Connor KE, Ramesh Babu P (2020) Bio-based and biodegradable polymers—state-of-the-art, challenges and emerging trends. Curr Opin Green Sustain Chem 21:75–81. https://doi.org/10.1016/j.cogsc.2019.12.005

Ravichandran R (2009) Nanoparticles in drug delivery: potential green nanobiomedicine applications. Int J Green Nanotechnol Biomed 1:108–130. https://doi.org/10.1080/19430850903430427

Rosentrater KA, Otieno AW (2006) Considerations for manufacturing bio-based plastic products. J Polym Environ. https://doi.org/10.1007/s10924-006-0036-1

Sadh PK, Duhan S, Duhan JS (2018) Agro-industrial wastes and their utilization using solid state fermentation: a review. Bioresour. Bioprocess.

Sangsuwan J, Rattanapanone N, Rachtanapun P (2008) Effect of chitosan/methyl cellulose films on microbial and quality characteristics of fresh-cut cantaloupe and pineapple. Postharvest Biol Technol 49:403–410. https://doi.org/10.1016/j.postharvbio.2008.02.014

Santana RF, Bonomo RCF, Gandolfi ORR et al (2018) Characterization of starch-based bioplastics from jackfruit seed plasticized with glycerol. J Food Sci Technol 55:278–286. https://doi.org/10.1007/s13197-017-2936-6

Schlenoff J, Decher G (2003) Multilayer thin films: sequential assembly of nanocomposite materials

Sharma P, Gaur VK, Kim S-H, Pandey A (2020) Microbial strategies for bio-transforming food waste into resources. Bioresour Technol 299. https://doi.org/10.1016/j.biortech.2019.122580.122580

Sinharay S, Bousmina M (2005) Biodegradable polymers and their layered silicate nanocomposites: in greening the 21st century materials world. Prog Mater Sci 50:962–1079. https://doi.org/10.1016/j.pmatsci.2005.05.002

Sozer N, Kokini JL (2009) Nanotechnology and its applications in the food sector. Trends Biotechnol 27:82–89. https://doi.org/10.1016/j.tibtech.2008.10.010

Stollman U, Johansson F, Leufven A (2000) A. Packaging and food quality. In: Man CMD (ed) Shelf-life evaluation of foods. Aspen Publishers, Gaithersbury MD, pp 42–56

Thunwall M, Kuthanova V, Boldizar A, Rigdahl M (2008) Film blowing of thermoplastic starch. Carbohydr Polym 71:583–590. https://doi.org/10.1016/j.carbpol.2007.07.001

Treinyte J, Bridziuviene D, Fataraite-Urboniene E et al (2018) Forestry wastes filled polymer composites for agricultural use. J Clean Prod 205:388–406. https://doi.org/10.1016/j.jclepro.2018.09.012

Tsang YF, Kumar V, Samadar P et al (2019) Production of bioplastic through food waste valorization. Environ Int 127:625–644. https://doi.org/10.1016/j.envint.2019.03.076

Vega-Castro O, Contreras-Calderon J, León E et al (2016) Characterization of a polyhydroxy-yalkanoate obtained from pineapple peel waste using Ralsthonia eutropha. J Biotechnol 231:232–238. https://doi.org/10.1016/j.jbiotec.2016.06.018

Vermeiren L, Devlieghere F, van Beest M et al (1999) Developments in the active packaging of foods. Trends Food Sci Technol 10:77–86. https://doi.org/10.1016/S0924-2244(99)00032-1

Vox G, Loisi RV, Blanco I et al (2016) Mapping of agriculture plastic waste. Agric Agric Sci Procedia 8:583–591. https://doi.org/10.1016/j.aaspro.2016.02.080

Vroman I, Tighzert L (2009) Biodegradable Polymers. Materials (basel) 2:307–344. https://doi.org/10.3390/ma2020307

Weisenburger DD (1993) Human health effects of agrichemical use. Hum Pathol 24:571–576. https://doi.org/10.1016/0046-8177(93)90234-8

Weiss J, Takhistov P, McClements DJ (2006) Functional materials in food nanotechnology. J Food Sci 71:107–116. https://doi.org/10.1111/j.1750-3841.2006.00195.x

Wiles DM, Scott G (2006) Polyolefins with controlled environmental degradability. Polym Degrad Stab 91:1581–1592. https://doi.org/10.1016/j.polymdegradstab.2005.09.010

Xie Y, Niu X, Yang J et al (2020) Active biodegradable films based on the whole potato peel incorporated with bacterial cellulose and curcumin. Int J Biol Macromol 150:480–491. https://doi.org/10.1016/j.ijbiomac.2020.01.291

Yates MR, Barlow CY (2013) Life cycle assessments of biodegradable, commercial biopolymers —a critical review. Resour Conserv Recycl 78:54–66. https://doi.org/10.1016/j.resconrec.2013.06.010

Youssef AM, El-Sayed SM (2018) Bionanocomposites materials for food packaging applications: concepts and future outlook. Carbohydr Polym 193:19–27. https://doi.org/10.1016/j.carbpol.2018.03.088

Zhao Y, McDaniel M (2005) Sensory quality of foods associated with edible film and coating systems and shelf-life extension. In: Innovations in food packaging. Elsevier, pp 434–453

Printed in the United States
by Baker & Taylor Publisher Services